Microbiomes: A Very Short Introduction

VERY SHORT INTRODUCTIONS are for anyone wanting a stimulating and accessible way into a new subject. They are written by experts, and have been translated into more than 45 different languages.

The series began in 1995, and now covers a wide variety of topics in every discipline. The VSI library currently contains over 700 volumes—a Very Short Introduction to everything from Psychology and Philosophy of Science to American History and Relativity—and continues to grow in every subject area.

Very Short Introductions available now:

ABOLITIONISM Richard S. Newman
THE ABRAHAMIC RELIGIONS Charles L. Cohen
ACCOUNTING Christopher Nobes
ADOLESCENCE Peter K. Smith
THEODOR W. ADORNO Andrew Bowie
ADVERTISING Winston Fletcher
AERIAL WARFARE Frank Ledwidge
AESTHETICS Bence Nanay
AFRICAN AMERICAN RELIGION Eddie S. Glaude Jr
AFRICAN HISTORY John Parker and Richard Rathbone
AFRICAN POLITICS Ian Taylor
AFRICAN RELIGIONS Jacob K. Olupona
AGEING Nancy A. Pachana
AGNOSTICISM Robin Le Poidevin
AGRICULTURE Paul Brassley and Richard Soffe
ALEXANDER THE GREAT Hugh Bowden
ALGEBRA Peter M. Higgins
AMERICAN BUSINESS HISTORY Walter A. Friedman
AMERICAN CULTURAL HISTORY Eric Avila
AMERICAN FOREIGN RELATIONS Andrew Preston
AMERICAN HISTORY Paul S. Boyer
AMERICAN IMMIGRATION David A. Gerber
AMERICAN INTELLECTUAL HISTORY Jennifer Ratner-Rosenhagen
THE AMERICAN JUDICIAL SYSTEM Charles L. Zelden
AMERICAN LEGAL HISTORY G. Edward White
AMERICAN MILITARY HISTORY Joseph T. Glatthaar
AMERICAN NAVAL HISTORY Craig L. Symonds
AMERICAN POETRY David Caplan
AMERICAN POLITICAL HISTORY Donald Critchlow
AMERICAN POLITICAL PARTIES AND ELECTIONS L. Sandy Maisel
AMERICAN POLITICS Richard M. Valelly
THE AMERICAN PRESIDENCY Charles O. Jones
THE AMERICAN REVOLUTION Robert J. Allison
AMERICAN SLAVERY Heather Andrea Williams
THE AMERICAN SOUTH Charles Reagan Wilson
THE AMERICAN WEST Stephen Aron
AMERICAN WOMEN'S HISTORY Susan Ware
AMPHIBIANS T. S. Kemp
ANAESTHESIA Aidan O'Donnell
ANALYTIC PHILOSOPHY Michael Beaney
ANARCHISM Alex Prichard

ANCIENT ASSYRIA Karen Radner
ANCIENT EGYPT Ian Shaw
ANCIENT EGYPTIAN ART AND ARCHITECTURE Christina Riggs
ANCIENT GREECE Paul Cartledge
THE ANCIENT NEAR EAST Amanda H. Podany
ANCIENT PHILOSOPHY Julia Annas
ANCIENT WARFARE Harry Sidebottom
ANGELS David Albert Jones
ANGLICANISM Mark Chapman
THE ANGLO-SAXON AGE John Blair
ANIMAL BEHAVIOUR Tristram D. Wyatt
THE ANIMAL KINGDOM Peter Holland
ANIMAL RIGHTS David DeGrazia
THE ANTARCTIC Klaus Dodds
ANTHROPOCENE Erle C. Ellis
ANTISEMITISM Steven Beller
ANXIETY Daniel Freeman and Jason Freeman
THE APOCRYPHAL GOSPELS Paul Foster
APPLIED MATHEMATICS Alain Goriely
THOMAS AQUINAS Fergus Kerr
ARBITRATION Thomas Schultz and Thomas Grant
ARCHAEOLOGY Paul Bahn
ARCHITECTURE Andrew Ballantyne
THE ARCTIC Klaus Dodds and Jamie Woodward
HANNAH ARENDT Dana Villa
ARISTOCRACY William Doyle
ARISTOTLE Jonathan Barnes
ART HISTORY Dana Arnold
ART THEORY Cynthia Freeland
ARTIFICIAL INTELLIGENCE Margaret A. Boden
ASIAN AMERICAN HISTORY Madeline Y. Hsu
ASTROBIOLOGY David C. Catling
ASTROPHYSICS James Binney
ATHEISM Julian Baggini
THE ATMOSPHERE Paul I. Palmer
AUGUSTINE Henry Chadwick
JANE AUSTEN Tom Keymer
AUSTRALIA Kenneth Morgan
AUTISM Uta Frith
AUTOBIOGRAPHY Laura Marcus
THE AVANT GARDE David Cottington
THE AZTECS Davíd Carrasco
BABYLONIA Trevor Bryce
BACTERIA Sebastian G. B. Amyes
BANKING John Goddard and John O. S. Wilson
BARTHES Jonathan Culler
THE BEATS David Sterritt
BEAUTY Roger Scruton
LUDWIG VAN BEETHOVEN Mark Evan Bonds
BEHAVIOURAL ECONOMICS Michelle Baddeley
BESTSELLERS John Sutherland
THE BIBLE John Riches
BIBLICAL ARCHAEOLOGY Eric H. Cline
BIG DATA Dawn E. Holmes
BIOCHEMISTRY Mark Lorch
BIOGEOGRAPHY Mark V. Lomolino
BIOGRAPHY Hermione Lee
BIOMETRICS Michael Fairhurst
ELIZABETH BISHOP Jonathan F. S. Post
BLACK HOLES Katherine Blundell
BLASPHEMY Yvonne Sherwood
BLOOD Chris Cooper
THE BLUES Elijah Wald
THE BODY Chris Shilling
NIELS BOHR J. L. Heilbron
THE BOOK OF COMMON PRAYER Brian Cummings
THE BOOK OF MORMON Terryl Givens
BORDERS Alexander C. Diener and Joshua Hagen
THE BRAIN Michael O'Shea
BRANDING Robert Jones
THE BRICS Andrew F. Cooper
THE BRITISH CONSTITUTION Martin Loughlin
THE BRITISH EMPIRE Ashley Jackson
BRITISH POLITICS Tony Wright
BUDDHA Michael Carrithers
BUDDHISM Damien Keown
BUDDHIST ETHICS Damien Keown
BYZANTIUM Peter Sarris
CALVINISM Jon Balserak

ALBERT CAMUS Oliver Gloag
CANADA Donald Wright
CANCER Nicholas James
CAPITALISM James Fulcher
CATHOLICISM Gerald O'Collins
CAUSATION Stephen Mumford and Rani Lill Anjum
THE CELL Terence Allen and Graham Cowling
THE CELTS Barry Cunliffe
CHAOS Leonard Smith
GEOFFREY CHAUCER David Wallace
CHEMISTRY Peter Atkins
CHILD PSYCHOLOGY Usha Goswami
CHILDREN'S LITERATURE Kimberley Reynolds
CHINESE LITERATURE Sabina Knight
CHOICE THEORY Michael Allingham
CHRISTIAN ART Beth Williamson
CHRISTIAN ETHICS D. Stephen Long
CHRISTIANITY Linda Woodhead
CIRCADIAN RHYTHMS Russell Foster and Leon Kreitzman
CITIZENSHIP Richard Bellamy
CITY PLANNING Carl Abbott
CIVIL ENGINEERING David Muir Wood
CLASSICAL LITERATURE William Allan
CLASSICAL MYTHOLOGY Helen Morales
CLASSICS Mary Beard and John Henderson
CLAUSEWITZ Michael Howard
CLIMATE Mark Maslin
CLIMATE CHANGE Mark Maslin
CLINICAL PSYCHOLOGY Susan Llewelyn and Katie Aafjes-van Doorn
COGNITIVE BEHAVIOURAL THERAPY Freda McManus
COGNITIVE NEUROSCIENCE Richard Passingham
THE COLD WAR Robert J. McMahon
COLONIAL AMERICA Alan Taylor
COLONIAL LATIN AMERICAN LITERATURE Rolena Adorno
COMBINATORICS Robin Wilson
COMEDY Matthew Bevis
COMMUNISM Leslie Holmes
COMPARATIVE LITERATURE Ben Hutchinson
COMPETITION AND ANTITRUST LAW Ariel Ezrachi
COMPLEXITY John H. Holland
THE COMPUTER Darrel Ince
COMPUTER SCIENCE Subrata Dasgupta
CONCENTRATION CAMPS Dan Stone
CONFUCIANISM Daniel K. Gardner
THE CONQUISTADORS Matthew Restall and Felipe Fernández-Armesto
CONSCIENCE Paul Strohm
CONSCIOUSNESS Susan Blackmore
CONTEMPORARY ART Julian Stallabrass
CONTEMPORARY FICTION Robert Eaglestone
CONTINENTAL PHILOSOPHY Simon Critchley
COPERNICUS Owen Gingerich
CORAL REEFS Charles Sheppard
CORPORATE SOCIAL RESPONSIBILITY Jeremy Moon
CORRUPTION Leslie Holmes
COSMOLOGY Peter Coles
COUNTRY MUSIC Richard Carlin
CREATIVITY Vlad Glăveanu
CRIME FICTION Richard Bradford
CRIMINAL JUSTICE Julian V. Roberts
CRIMINOLOGY Tim Newburn
CRITICAL THEORY Stephen Eric Bronner
THE CRUSADES Christopher Tyerman
CRYPTOGRAPHY Fred Piper and Sean Murphy
CRYSTALLOGRAPHY A. M. Glazer
THE CULTURAL REVOLUTION Richard Curt Kraus
DADA AND SURREALISM David Hopkins
DANTE Peter Hainsworth and David Robey
DARWIN Jonathan Howard
THE DEAD SEA SCROLLS Timothy H. Lim
DECADENCE David Weir
DECOLONIZATION Dane Kennedy

DEMENTIA Kathleen Taylor
DEMOCRACY Bernard Crick
DEMOGRAPHY Sarah Harper
DEPRESSION Jan Scott and Mary Jane Tacchi
DERRIDA Simon Glendinning
DESCARTES Tom Sorell
DESERTS Nick Middleton
DESIGN John Heskett
DEVELOPMENT Ian Goldin
DEVELOPMENTAL BIOLOGY Lewis Wolpert
THE DEVIL Darren Oldridge
DIASPORA Kevin Kenny
CHARLES DICKENS Jenny Hartley
DICTIONARIES Lynda Mugglestone
DINOSAURS David Norman
DIPLOMATIC HISTORY Joseph M. Siracusa
DOCUMENTARY FILM Patricia Aufderheide
DREAMING J. Allan Hobson
DRUGS Les Iversen
DRUIDS Barry Cunliffe
DYNASTY Jeroen Duindam
DYSLEXIA Margaret J. Snowling
EARLY MUSIC Thomas Forrest Kelly
THE EARTH Martin Redfern
EARTH SYSTEM SCIENCE Tim Lenton
ECOLOGY Jaboury Ghazoul
ECONOMICS Partha Dasgupta
EDUCATION Gary Thomas
EGYPTIAN MYTH Geraldine Pinch
EIGHTEENTH-CENTURY BRITAIN Paul Langford
THE ELEMENTS Philip Ball
EMOTION Dylan Evans
EMPIRE Stephen Howe
EMPLOYMENT LAW David Cabrelli
ENERGY SYSTEMS Nick Jenkins
ENGELS Terrell Carver
ENGINEERING David Blockley
THE ENGLISH LANGUAGE Simon Horobin
ENGLISH LITERATURE Jonathan Bate
THE ENLIGHTENMENT John Robertson
ENTREPRENEURSHIP Paul Westhead and Mike Wright
ENVIRONMENTAL ECONOMICS Stephen Smith
ENVIRONMENTAL ETHICS Robin Attfield
ENVIRONMENTAL LAW Elizabeth Fisher
ENVIRONMENTAL POLITICS Andrew Dobson
ENZYMES Paul Engel
EPICUREANISM Catherine Wilson
EPIDEMIOLOGY Rodolfo Saracci
ETHICS Simon Blackburn
ETHNOMUSICOLOGY Timothy Rice
THE ETRUSCANS Christopher Smith
EUGENICS Philippa Levine
THE EUROPEAN UNION Simon Usherwood and John Pinder
EUROPEAN UNION LAW Anthony Arnull
EVANGELICALISM John G. Stackhouse Jr.
EVIL Luke Russell
EVOLUTION Brian and Deborah Charlesworth
EXISTENTIALISM Thomas Flynn
EXPLORATION Stewart A. Weaver
EXTINCTION Paul B. Wignall
THE EYE Michael Land
FAIRY TALE Marina Warner
FAMILY LAW Jonathan Herring
MICHAEL FARADAY Frank A. J. L. James
FASCISM Kevin Passmore
FASHION Rebecca Arnold
FEDERALISM Mark J. Rozell and Clyde Wilcox
FEMINISM Margaret Walters
FILM Michael Wood
FILM MUSIC Kathryn Kalinak
FILM NOIR James Naremore
FIRE Andrew C. Scott
THE FIRST WORLD WAR Michael Howard
FLUID MECHANICS Eric Lauga
FOLK MUSIC Mark Slobin
FOOD John Krebs
FORENSIC PSYCHOLOGY David Canter
FORENSIC SCIENCE Jim Fraser
FORESTS Jaboury Ghazoul

FOSSILS Keith Thomson
FOUCAULT Gary Gutting
THE FOUNDING FATHERS
R. B. Bernstein
FRACTALS Kenneth Falconer
FREE SPEECH Nigel Warburton
FREE WILL Thomas Pink
FREEMASONRY Andreas Önnerfors
FRENCH LITERATURE
John D. Lyons
FRENCH PHILOSOPHY
Stephen Gaukroger and Knox Peden
THE FRENCH REVOLUTION
William Doyle
FREUD Anthony Storr
FUNDAMENTALISM Malise Ruthven
FUNGI Nicholas P. Money
THE FUTURE Jennifer M. Gidley
GALAXIES John Gribbin
GALILEO Stillman Drake
GAME THEORY Ken Binmore
GANDHI Bhikhu Parekh
GARDEN HISTORY Gordon Campbell
GENES Jonathan Slack
GENIUS Andrew Robinson
GENOMICS John Archibald
GEOGRAPHY John Matthews and
David Herbert
GEOLOGY Jan Zalasiewicz
GEOMETRY Maciej Dunajski
GEOPHYSICS William Lowrie
GEOPOLITICS Klaus Dodds
GERMAN LITERATURE
Nicholas Boyle
GERMAN PHILOSOPHY
Andrew Bowie
THE GHETTO Bryan Cheyette
GLACIATION David J. A. Evans
GLOBAL CATASTROPHES Bill McGuire
GLOBAL ECONOMIC HISTORY
Robert C. Allen
GLOBAL ISLAM Nile Green
GLOBALIZATION Manfred B. Steger
GOD John Bowker
GÖDEL'S THEOREM A. W. Moore
GOETHE Ritchie Robertson
THE GOTHIC Nick Groom
GOVERNANCE Mark Bevir
GRAVITY Timothy Clifton
THE GREAT DEPRESSION AND
THE NEW DEAL Eric Rauchway
HABEAS CORPUS Amanda L. Tyler
HABERMAS James Gordon Finlayson
THE HABSBURG EMPIRE
Martyn Rady
HAPPINESS Daniel M. Haybron
THE HARLEM RENAISSANCE
Cheryl A. Wall
THE HEBREW BIBLE AS
LITERATURE Tod Linafelt
HEGEL Peter Singer
HEIDEGGER Michael Inwood
THE HELLENISTIC AGE
Peter Thonemann
HEREDITY John Waller
HERMENEUTICS Jens Zimmermann
HERODOTUS Jennifer T. Roberts
HIEROGLYPHS Penelope Wilson
HINDUISM Kim Knott
HISTORY John H. Arnold
THE HISTORY OF ASTRONOMY
Michael Hoskin
THE HISTORY OF CHEMISTRY
William H. Brock
THE HISTORY OF CHILDHOOD
James Marten
THE HISTORY OF CINEMA
Geoffrey Nowell-Smith
THE HISTORY OF COMPUTING
Doron Swade
THE HISTORY OF LIFE Michael Benton
THE HISTORY OF MATHEMATICS
Jacqueline Stedall
THE HISTORY OF MEDICINE
William Bynum
THE HISTORY OF PHYSICS
J. L. Heilbron
THE HISTORY OF POLITICAL
THOUGHT Richard Whatmore
THE HISTORY OF TIME
Leofranc Holford-Strevens
HIV AND AIDS Alan Whiteside
HOBBES Richard Tuck
HOLLYWOOD Peter Decherney
THE HOLY ROMAN EMPIRE
Joachim Whaley
HOME Michael Allen Fox
HOMER Barbara Graziosi
HORMONES Martin Luck
HORROR Darryl Jones
HUMAN ANATOMY Leslie Klenerman
HUMAN EVOLUTION Bernard Wood

HUMAN PHYSIOLOGY
Jamie A. Davies
HUMAN RESOURCE MANAGEMENT
Adrian Wilkinson
HUMAN RIGHTS Andrew Clapham
HUMANISM Stephen Law
HUME James A. Harris
HUMOUR Noël Carroll
THE ICE AGE Jamie Woodward
IDENTITY Florian Coulmas
IDEOLOGY Michael Freeden
THE IMMUNE SYSTEM
Paul Klenerman
INDIAN CINEMA
Ashish Rajadhyaksha
INDIAN PHILOSOPHY Sue Hamilton
THE INDUSTRIAL REVOLUTION
Robert C. Allen
INFECTIOUS DISEASE Marta L. Wayne and Benjamin M. Bolker
INFINITY Ian Stewart
INFORMATION Luciano Floridi
INNOVATION Mark Dodgson and David Gann
INTELLECTUAL PROPERTY
Siva Vaidhyanathan
INTELLIGENCE Ian J. Deary
INTERNATIONAL LAW
Vaughan Lowe
INTERNATIONAL MIGRATION
Khalid Koser
INTERNATIONAL RELATIONS
Christian Reus-Smit
INTERNATIONAL SECURITY
Christopher S. Browning
INSECTS Simon Leather
IRAN Ali M. Ansari
ISLAM Malise Ruthven
ISLAMIC HISTORY Adam Silverstein
ISLAMIC LAW Mashood A. Baderin
ISOTOPES Rob Ellam
ITALIAN LITERATURE
Peter Hainsworth and David Robey
HENRY JAMES Susan L. Mizruchi
JESUS Richard Bauckham
JEWISH HISTORY David N. Myers
JEWISH LITERATURE Ilan Stavans
JOURNALISM Ian Hargreaves
JAMES JOYCE Colin MacCabe
JUDAISM Norman Solomon
JUNG Anthony Stevens
KABBALAH Joseph Dan
KAFKA Ritchie Robertson
KANT Roger Scruton
KEYNES Robert Skidelsky
KIERKEGAARD Patrick Gardiner
KNOWLEDGE Jennifer Nagel
THE KORAN Michael Cook
KOREA Michael J. Seth
LAKES Warwick F. Vincent
LANDSCAPE ARCHITECTURE
Ian H. Thompson
LANDSCAPES AND GEOMORPHOLOGY
Andrew Goudie and Heather Viles
LANGUAGES Stephen R. Anderson
LATE ANTIQUITY Gillian Clark
LAW Raymond Wacks
THE LAWS OF THERMODYNAMICS
Peter Atkins
LEADERSHIP Keith Grint
LEARNING Mark Haselgrove
LEIBNIZ Maria Rosa Antognazza
C. S. LEWIS James Como
LIBERALISM Michael Freeden
LIGHT Ian Walmsley
LINCOLN Allen C. Guelzo
LINGUISTICS Peter Matthews
LITERARY THEORY Jonathan Culler
LOCKE John Dunn
LOGIC Graham Priest
LOVE Ronald de Sousa
MARTIN LUTHER Scott H. Hendrix
MACHIAVELLI Quentin Skinner
MADNESS Andrew Scull
MAGIC Owen Davies
MAGNA CARTA Nicholas Vincent
MAGNETISM Stephen Blundell
MALTHUS Donald Winch
MAMMALS T. S. Kemp
MANAGEMENT John Hendry
NELSON MANDELA Elleke Boehmer
MAO Delia Davin
MARINE BIOLOGY Philip V. Mladenov
MARKETING
Kenneth Le Meunier-FitzHugh
THE MARQUIS DE SADE John Phillips
MARTYRDOM Jolyon Mitchell
MARX Peter Singer
MATERIALS Christopher Hall
MATHEMATICAL FINANCE
Mark H. A. Davis

MATHEMATICS Timothy Gowers
MATTER Geoff Cottrell
THE MAYA Matthew Restall and Amara Solari
THE MEANING OF LIFE Terry Eagleton
MEASUREMENT David Hand
MEDICAL ETHICS Michael Dunn and Tony Hope
MEDICAL LAW Charles Foster
MEDIEVAL BRITAIN John Gillingham and Ralph A. Griffiths
MEDIEVAL LITERATURE Elaine Treharne
MEDIEVAL PHILOSOPHY John Marenbon
MEMORY Jonathan K. Foster
METAPHYSICS Stephen Mumford
METHODISM William J. Abraham
THE MEXICAN REVOLUTION Alan Knight
MICROBIOLOGY Nicholas P. Money
MICROBIOMES Angela E. Douglas
MICROECONOMICS Avinash Dixit
MICROSCOPY Terence Allen
THE MIDDLE AGES Miri Rubin
MILITARY JUSTICE Eugene R. Fidell
MILITARY STRATEGY Antulio J. Echevarria II
JOHN STUART MILL Gregory Claeys
MINERALS David Vaughan
MIRACLES Yujin Nagasawa
MODERN ARCHITECTURE Adam Sharr
MODERN ART David Cottington
MODERN BRAZIL Anthony W. Pereira
MODERN CHINA Rana Mitter
MODERN DRAMA Kirsten E. Shepherd-Barr
MODERN FRANCE Vanessa R. Schwartz
MODERN INDIA Craig Jeffrey
MODERN IRELAND Senia Pašeta
MODERN ITALY Anna Cento Bull
MODERN JAPAN Christopher Goto-Jones
MODERN LATIN AMERICAN LITERATURE Roberto González Echevarría
MODERN WAR Richard English
MODERNISM Christopher Butler
MOLECULAR BIOLOGY Aysha Divan and Janice A. Royds
MOLECULES Philip Ball
MONASTICISM Stephen J. Davis
THE MONGOLS Morris Rossabi
MONTAIGNE William M. Hamlin
MOONS David A. Rothery
MORMONISM Richard Lyman Bushman
MOUNTAINS Martin F. Price
MUHAMMAD Jonathan A. C. Brown
MULTICULTURALISM Ali Rattansi
MULTILINGUALISM John C. Maher
MUSIC Nicholas Cook
MUSIC AND TECHNOLOGY Mark Katz
MYTH Robert A. Segal
NAPOLEON David A. Bell
THE NAPOLEONIC WARS Mike Rapport
NATIONALISM Steven Grosby
NATIVE AMERICAN LITERATURE Sean Teuton
NAVIGATION Jim Bennett
NAZI GERMANY Jane Caplan
NEGOTIATION Carrie Menkel-Meadow
NEOLIBERALISM Manfred B. Steger and Ravi K. Roy
NETWORKS Guido Caldarelli and Michele Catanzaro
THE NEW TESTAMENT Luke Timothy Johnson
THE NEW TESTAMENT AS LITERATURE Kyle Keefer
NEWTON Robert Iliffe
NIETZSCHE Michael Tanner
NINETEENTH-CENTURY BRITAIN Christopher Harvie and H. C. G. Matthew
THE NORMAN CONQUEST George Garnett
NORTH AMERICAN INDIANS Theda Perdue and Michael D. Green
NORTHERN IRELAND Marc Mulholland
NOTHING Frank Close
NUCLEAR PHYSICS Frank Close
NUCLEAR POWER Maxwell Irvine
NUCLEAR WEAPONS Joseph M. Siracusa

NUMBER THEORY Robin Wilson
NUMBERS Peter M. Higgins
NUTRITION David A. Bender
OBJECTIVITY Stephen Gaukroger
OCEANS Dorrik Stow
THE OLD TESTAMENT
Michael D. Coogan
THE ORCHESTRA D. Kern Holoman
ORGANIC CHEMISTRY
Graham Patrick
ORGANIZATIONS Mary Jo Hatch
ORGANIZED CRIME
Georgios A. Antonopoulos and
Georgios Papanicolaou
ORTHODOX CHRISTIANITY
A. Edward Siecienski
OVID Llewelyn Morgan
PAGANISM Owen Davies
PAKISTAN Pippa Virdee
THE PALESTINIAN-ISRAELI
CONFLICT Martin Bunton
PANDEMICS Christian W. McMillen
PARTICLE PHYSICS Frank Close
PAUL E. P. Sanders
IVAN PAVLOV Daniel P. Todes
PEACE Oliver P. Richmond
PENTECOSTALISM William K. Kay
PERCEPTION Brian Rogers
THE PERIODIC TABLE Eric R. Scerri
PHILOSOPHICAL METHOD
Timothy Williamson
PHILOSOPHY Edward Craig
PHILOSOPHY IN THE ISLAMIC
WORLD Peter Adamson
PHILOSOPHY OF BIOLOGY
Samir Okasha
PHILOSOPHY OF LAW
Raymond Wacks
PHILOSOPHY OF MIND
Barbara Gail Montero
PHILOSOPHY OF PHYSICS
David Wallace
PHILOSOPHY OF SCIENCE
Samir Okasha
PHILOSOPHY OF RELIGION
Tim Bayne
PHOTOGRAPHY Steve Edwards
PHYSICAL CHEMISTRY Peter Atkins
PHYSICS Sidney Perkowitz
PILGRIMAGE Ian Reader
PLAGUE Paul Slack
PLANETARY SYSTEMS
Raymond T. Pierrehumbert
PLANETS David A. Rothery
PLANTS Timothy Walker
PLATE TECTONICS Peter Molnar
PLATO Julia Annas
POETRY Bernard O'Donoghue
POLITICAL PHILOSOPHY
David Miller
POLITICS Kenneth Minogue
POLYGAMY Sarah M. S. Pearsall
POPULISM Cas Mudde and
Cristóbal Rovira Kaltwasser
POSTCOLONIALISM Robert J. C. Young
POSTMODERNISM Christopher Butler
POSTSTRUCTURALISM
Catherine Belsey
POVERTY Philip N. Jefferson
PREHISTORY Chris Gosden
PRESOCRATIC
PHILOSOPHY Catherine Osborne
PRIVACY Raymond Wacks
PROBABILITY John Haigh
PROGRESSIVISM Walter Nugent
PROHIBITION W. J. Rorabaugh
PROJECTS Andrew Davies
PROTESTANTISM Mark A. Noll
PSYCHIATRY Tom Burns
PSYCHOANALYSIS Daniel Pick
PSYCHOLOGY Gillian Butler and
Freda McManus
PSYCHOLOGY OF MUSIC
Elizabeth Hellmuth Margulis
PSYCHOPATHY Essi Viding
PSYCHOTHERAPY Tom Burns and
Eva Burns-Lundgren
PUBLIC ADMINISTRATION
Stella Z. Theodoulou and Ravi K. Roy
PUBLIC HEALTH Virginia Berridge
PURITANISM Francis J. Bremer
THE QUAKERS Pink Dandelion
QUANTUM THEORY
John Polkinghorne
RACISM Ali Rattansi
RADIOACTIVITY Claudio Tuniz
RASTAFARI Ennis B. Edmonds
READING Belinda Jack
THE REAGAN REVOLUTION Gil Troy
REALITY Jan Westerhoff
RECONSTRUCTION Allen C. Guelzo
THE REFORMATION Peter Marshall

REFUGEES Gil Loescher
RELATIVITY Russell Stannard
RELIGION Thomas A. Tweed
RELIGION IN AMERICA Timothy Beal
THE RENAISSANCE Jerry Brotton
RENAISSANCE ART Geraldine A. Johnson
RENEWABLE ENERGY Nick Jelley
REPTILES T. S. Kemp
REVOLUTIONS Jack A. Goldstone
RHETORIC Richard Toye
RISK Baruch Fischhoff and John Kadvany
RITUAL Barry Stephenson
RIVERS Nick Middleton
ROBOTICS Alan Winfield
ROCKS Jan Zalasiewicz
ROMAN BRITAIN Peter Salway
THE ROMAN EMPIRE Christopher Kelly
THE ROMAN REPUBLIC David M. Gwynn
ROMANTICISM Michael Ferber
ROUSSEAU Robert Wokler
RUSSELL A. C. Grayling
THE RUSSIAN ECONOMY Richard Connolly
RUSSIAN HISTORY Geoffrey Hosking
RUSSIAN LITERATURE Catriona Kelly
THE RUSSIAN REVOLUTION S. A. Smith
SAINTS Simon Yarrow
SAMURAI Michael Wert
SAVANNAS Peter A. Furley
SCEPTICISM Duncan Pritchard
SCHIZOPHRENIA Chris Frith and Eve Johnstone
SCHOPENHAUER Christopher Janaway
SCIENCE AND RELIGION Thomas Dixon and Adam R. Shapiro
SCIENCE FICTION David Seed
THE SCIENTIFIC REVOLUTION Lawrence M. Principe
SCOTLAND Rab Houston
SECULARISM Andrew Copson
SEXUAL SELECTION Marlene Zuk and Leigh W. Simmons
SEXUALITY Véronique Mottier
WILLIAM SHAKESPEARE Stanley Wells
SHAKESPEARE'S COMEDIES Bart van Es
SHAKESPEARE'S SONNETS AND POEMS Jonathan F. S. Post
SHAKESPEARE'S TRAGEDIES Stanley Wells
GEORGE BERNARD SHAW Christopher Wixson
MARY SHELLEY Charlotte Gordon
THE SHORT STORY Andrew Kahn
SIKHISM Eleanor Nesbitt
SILENT FILM Donna Kornhaber
THE SILK ROAD James A. Millward
SLANG Jonathon Green
SLEEP Steven W. Lockley and Russell G. Foster
SMELL Matthew Cobb
ADAM SMITH Christopher J. Berry
SOCIAL AND CULTURAL ANTHROPOLOGY John Monaghan and Peter Just
SOCIAL PSYCHOLOGY Richard J. Crisp
SOCIAL WORK Sally Holland and Jonathan Scourfield
SOCIALISM Michael Newman
SOCIOLINGUISTICS John Edwards
SOCIOLOGY Steve Bruce
SOCRATES C. C. W. Taylor
SOFT MATTER Tom McLeish
SOUND Mike Goldsmith
SOUTHEAST ASIA James R. Rush
THE SOVIET UNION Stephen Lovell
THE SPANISH CIVIL WAR Helen Graham
SPANISH LITERATURE Jo Labanyi
THE SPARTANS Andrew J. Bayliss
SPINOZA Roger Scruton
SPIRITUALITY Philip Sheldrake
SPORT Mike Cronin
STARS Andrew King
STATISTICS David J. Hand
STEM CELLS Jonathan Slack
STOICISM Brad Inwood
STRUCTURAL ENGINEERING David Blockley
STUART BRITAIN John Morrill
THE SUN Philip Judge
SUPERCONDUCTIVITY Stephen Blundell
SUPERSTITION Stuart Vyse

SYMMETRY Ian Stewart
SYNAESTHESIA Julia Simner
SYNTHETIC BIOLOGY Jamie A. Davies
SYSTEMS BIOLOGY Eberhard O. Voit
TAXATION Stephen Smith
TEETH Peter S. Ungar
TELESCOPES Geoff Cottrell
TERRORISM Charles Townshend
THEATRE Marvin Carlson
THEOLOGY David F. Ford
THINKING AND REASONING Jonathan St B. T. Evans
THOUGHT Tim Bayne
TIBETAN BUDDHISM Matthew T. Kapstein
TIDES David George Bowers and Emyr Martyn Roberts
TIME Jenann Ismael
TOCQUEVILLE Harvey C. Mansfield
LEO TOLSTOY Liza Knapp
TOPOLOGY Richard Earl
TRAGEDY Adrian Poole
TRANSLATION Matthew Reynolds
THE TREATY OF VERSAILLES Michael S. Neiberg
TRIGONOMETRY Glen Van Brummelen
THE TROJAN WAR Eric H. Cline
TRUST Katherine Hawley
THE TUDORS John Guy
TWENTIETH-CENTURY BRITAIN Kenneth O. Morgan
TYPOGRAPHY Paul Luna
THE UNITED NATIONS Jussi M. Hanhimäki
UNIVERSITIES AND COLLEGES David Palfreyman and Paul Temple
THE U.S. CIVIL WAR Louis P. Masur
THE U.S. CONGRESS Donald A. Ritchie
THE U.S. CONSTITUTION David J. Bodenhamer
THE U.S. SUPREME COURT Linda Greenhouse
UTILITARIANISM Katarzyna de Lazari-Radek and Peter Singer
UTOPIANISM Lyman Tower Sargent
VETERINARY SCIENCE James Yeates
THE VIKINGS Julian D. Richards
VIOLENCE Philip Dwyer
THE VIRGIN MARY Mary Joan Winn Leith
THE VIRTUES Craig A. Boyd and Kevin Timpe
VIRUSES Dorothy H. Crawford
VOLCANOES Michael J. Branney and Jan Zalasiewicz
VOLTAIRE Nicholas Cronk
WAR AND RELIGION Jolyon Mitchell and Joshua Rey
WAR AND TECHNOLOGY Alex Roland
WATER John Finney
WAVES Mike Goldsmith
WEATHER Storm Dunlop
THE WELFARE STATE David Garland
WITCHCRAFT Malcolm Gaskill
WITTGENSTEIN A. C. Grayling
WORK Stephen Fineman
WORLD MUSIC Philip Bohlman
WORLD MYTHOLOGY David Leeming
THE WORLD TRADE ORGANIZATION Amrita Narlikar
WORLD WAR II Gerhard L. Weinberg
WRITING AND SCRIPT Andrew Robinson
ZIONISM Michael Stanislawski
ÉMILE ZOLA Brian Nelson

Available soon:

ANSELM Thomas Williams

For more information visit our website

www.oup.com/vsi/

Angela E. Douglas

MICROBIOMES

A Very Short Introduction

Great Clarendon Street, Oxford, OX2 6DP,
United Kingdom

Oxford University Press is a department of the University of Oxford.
It furthers the University's objective of excellence in research, scholarship,
and education by publishing worldwide. Oxford is a registered trade mark of
Oxford University Press in the UK and in certain other countries

First edition published in 2022

Impression: 2

Published in the United States of America by Oxford University Press
198 Madison Avenue, New York, NY 10016, United States of America

British Library Cataloguing in Publication Data
Data available

Library of Congress Control Number: 2022939623

ISBN 978–0–19–887085–2

Printed and bound by
CPI Group (UK) Ltd, Croydon, CR0 4YY

Contents

Preface

The term microbiome is a scientific outlier. It is one of the few terms adopted by scientists that has broken out of the scientific literature into everyday use. In 2001, Joshua Lederberg and Alexa McCray coined the term, and microbiome was swiftly adopted by their scientific colleagues to describe the community of microorganisms associated with humans, other animals, and plants. At that time, no one would have predicted that 'microbiome' would come to be a favourite buzzword in a diversity of wellness magazines and websites. Many of these publications are informative and interesting, but some include serious scientific errors. All too often, we are encouraged to attribute our physical and mental health (or ill-health) to our microbiome; exhorted to modify our diet or habits to improve our microbiome; and instructed to purchase certain products that will enhance our microbiome and, thereby, guarantee our health and happiness.

Several years ago, I recognized that there was a need for a straightforward guide to the science of microbiomes that summarizes the core concepts and facts. This information provides the basis for readers to sift fact and rational expectation from fiction and bogus claims. I am delighted that Oxford University Press has given me the opportunity to write this *Very Short Introduction* on Microbiomes.

Today, most of the research on microbiomes is biomedical, meaning that it is focused on human health and disease. I devote three chapters (Chapters 3–5) to the impact of microbiomes on human metabolic health, mental health, and susceptibility to infectious disease. Much of our understanding of these effects comes from observations and experiments conducted on non-human animals. These studies are very relevant to humans for the simple reason that the human microbiome functions and has been acquired in ways that parallel the microbiomes in many other animals and even plants. I explore these topics in Chapters 1 and 2. Chapter 6 concerns the microbiomes of plants, with an emphasis on how the microbiomes can contribute to sustainable agricultural practices and improved food production. In the final chapter, Chapter 7, I return to human health. In the light of growing evidence that the human microbiome is being depleted by modern lifestyles and the overuse of antimicrobials, I consider realistic strategies and therapies that can be applied to restore health-promoting microbiomes.

I have written this book to be enjoyed by any reader with a basic understanding of biology; I do not assume university-level scientific education. To assist readers, I provide a Glossary of key biological terms and a short list of books and articles (Further reading) which offer in-depth discussions of topics introduced in this book.

I am very grateful to Latha Menon, who commissioned this book and provided me with helpful advice and encouragement throughout its gestation. I also thank Dan Buckley, Maria Fernandez, Maria Harrison, Helene Marquis, Jeremy Searle, Ken Searle, and Mark Searle who generously read early drafts. Their comments and recommendations have made all the difference. I am responsible for any inaccuracies that remain.

Angela E. Douglas
29 March 2022

List of illustrations

List of tables

Chapter 1
Living with microbes

We like to think that we are human, but that is only partly true. Half of the estimated 80 trillion cells in the human body are microorganisms, and these microorganisms are crucial for our health and wellbeing. They protect us from pathogens, support our metabolism and immune system, and influence our mood and emotional state. Furthermore, bearing microorganisms is not special to humans. All healthy animals and plants are inhabited by communities of microorganisms known as microbiomes, and, generally, these microorganisms contribute to good health.

In the beginning, there were bacteria

The bacteria that dominate our microbiomes are the direct descendants of the first single-celled and morphologically simple organisms which evolved within a hundred million years after the planet became cool and wet enough for life. This was about 3.6 billion years ago (Figure 1(a)). Today, bacteria occupy every habitat on the planet and, despite their morphological simplicity, they have a rich social life. It is commonplace for bacteria to share resources and cooperate with other members of the same and different species; bacteria have sex; and bacteria can engage in the most ferocious biological warfare.

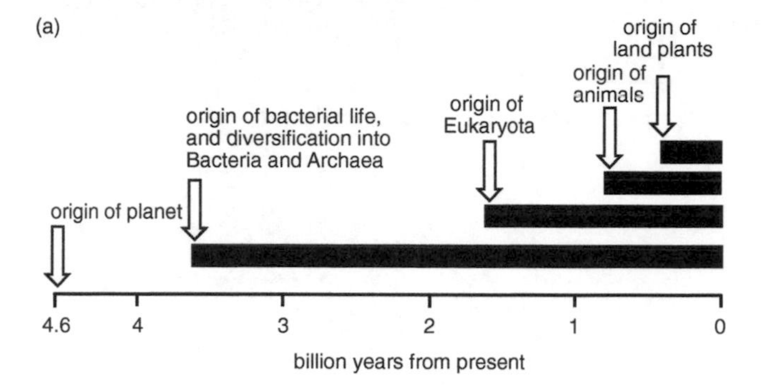

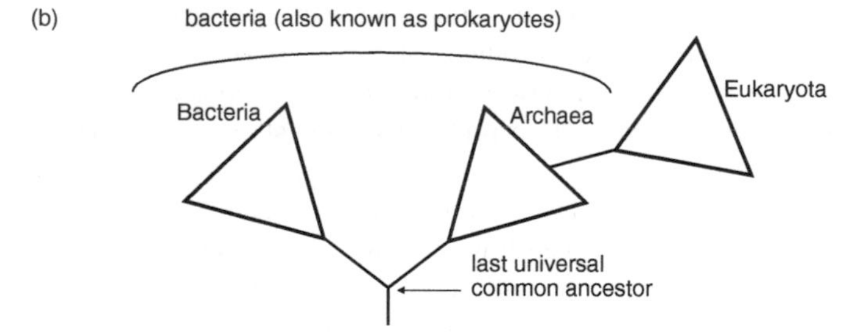

1. The origin and diversity of living organisms. (a) Evolutionary history of life on planet Earth. A billion years is one thousand million years. (b) The bacteria comprise two domains, the Bacteria and the Archaea. (The capital 'B' of the domain Bacteria distinguishes this domain from bacteria as a grade of organization with a lower case 'b'.) The eukaryotes (domain Eukaryota) evolved from one group of Archaea, known as the Asgard Archaea. Most eukaryotes live as single cells but several groups, including the animals and plants, are multicellular (see box: Microorganisms).

The bacteria had the planet to themselves for approximately 2 billion years, and many biologists argue that bacteria remain the dominant living organisms to this day. Early in their evolutionary history, the bacteria diversified into two major groups, one called the Archaea and the other called the Bacteria (Figure 1(b)). As shown in this figure, the term 'bacteria' (lower case 'b') refers to a grade of biological organization, while 'Bacteria' (upper case 'B') refers to one of the two groups of bacteria.

The eukaryotes: latecomers in the history of life

About 1.6 billion years ago, a novel life form that we call the eukaryotes evolved from among the Archaea. The eukaryotes are morphologically more complex than other organisms, in two ways. Eukaryotic cells are larger and contain a nucleus and other multiple membrane-bound compartments, and eukaryotes are much more likely than bacteria to be multicellular, meaning that the organism is made up of many cells with diverse functions. Nevertheless, many eukaryotes today live as single cells. Together with the Bacteria and Archaea, these single-celled eukaryotes are referred to as microorganisms.

Microorganisms

The formal definition of a microorganism is any organism that cannot be seen with the naked eye, which means organisms of dimension less than 1 mm. Because most very small organisms comprise just a single cell, most biologists treat organisms which are unicellular for all or most of their life cycle as microorganisms, but do not describe tiny animals (e.g. some nematode worms and insects which comprise many cells but are <1 mm in length) as microorganisms.

The microorganisms comprise three groups: all members of the Bacteria and the Archaea, and small Eukaryota (see Figure 1). The Bacteria and Archaea differ from each other in many ways, but they are both known as bacteria (lower case 'b') because their genetic material (DNA) is not separated from the rest of the cell contents by a membrane-bound nucleus, unlike eukaryotic cells.

(*continued*)

(In some texts, the bacteria are referred to as prokaryotes.) Microbial eukaryotes are very diverse, and include amoebae, algae, unicellular fungi (known as yeasts), and small, multicellular fungi. Most multicellular eukaryotes, notably the animals, land plants, seaweeds, and large fungal structures (e.g. mushrooms), are not microorganisms. Most bacteria are 0.5–5.0 μm in size, while eukaryotic microorganisms are generally 10–100 μm, but there is much variation.

Viruses comprise genetic material (DNA or RNA), often within a protein coat or membrane. They are parasites of living cells (Bacteria, Archaea, or Eukaryota) and use the cellular machinery of their host to reproduce. Some authorities consider viruses as a further type of microorganism, but it is hotly debated whether viruses are alive and how they originated. Some viruses are probably genetic material that has escaped from bacteria or eukaryotic cells, while others might be much-reduced bacteria (although their relationship to all known bacteria is obscure). Viruses that attack bacteria are known as bacteriophage, often abbreviated to phage. Most virus particles are 0.02–0.4 μm in dimension. The community of viruses in a host is often referred to as the virome.

We can be confident that the first eukaryotes interacted with many bacteria in their environment. Some of the interactions would have been antagonistic, equivalent to the many modern microbial eukaryotes that feed on bacteria, as well as various bacteria that are virulent pathogens of eukaryotes. Other partnerships between the first eukaryotes and bacteria were beneficial. A prime example is provided by the origin of mitochondria, the structures in the cells of most modern eukaryotes that use oxygen for respiration, yielding large amounts of energy. The mitochondria are derived from a *Rickettsia*-like bacterium that came to inhabit the cells of early eukaryotes, to mutual benefit. If this partnership had not evolved, the eukaryotes would have been a minor group of energy-limited microorganisms restricted to oxygen-free sediments.

Two groups of multicellular eukaryotes have been particularly successful: the animals, which evolved from unicellular eukaryotes in the ocean about 0.8 billion years ago; and the land plants, which arose from freshwater algae about 0.4 billion years ago (Figure 1(a)). Just like their single-celled ancestors, these multicellular eukaryotes interact persistently with bacteria. These associations are very unequal in terms of body size and number, such that a single large animal or plant, often referred to as the host, harbours many small microorganisms which, collectively, represent the microbiome.

The microbiomes of animals and plants

The microbiomes of many animals are predominantly found in the gut. For example, most of the microbial partners of humans are in one region of the gut, the colon (also known as the large intestine). The skin and upper airways are also extensively colonized, as is the vagina of women, but the internal organs (e.g. the liver and brain) bear few or no microorganisms in healthy people. The total weight of microorganisms in an adult human is estimated at 2–6 lb (1–3.7 kg), which is approximately 3 per cent of body weight; for comparison, the adult brain or liver each weigh 3 lb (1.4 kg). (In the opening paragraph of this chapter, I described how approximately half of the cells in our bodies are microbial and mostly bacteria. Microorganisms do not account for half of our weight because most microbial cells are much smaller than human cells.)

Some animals have highly specialized microbiomes that are housed in dedicated structures, known as symbiotic organs. For example, the symbiotic organs below the eyes of flashlight fish (Anomalopidae) are cavities densely colonized with luminescent bacteria. The fish communicate by precisely timed sequences of light flashes, as they raise and lower a dark shutter over their luminescent bacteria. In giant tube worms living in the hot, sulphide-rich waters that spew from deep sea hydrothermal vents, the gut is obliterated by a new organ, the trophosome, containing large numbers of bacteria which provide nutrients to the worm. Specialized symbiotic organs are also widespread in insects. For example, bedbugs and lice thrive on a nutrient-deficient diet of blood only because they obtain essential vitamins from single species of bacteria packed into an organ in their body cavity.

Plants are also host to a diversity of microorganisms, many of which colonize the surface or penetrate into the internal tissues of roots, leaves, and other plant parts. In natural soils, a dense community of microorganisms in contact with the root surface promotes plant growth and provides protection against pathogens. Although many of the microbial partners of plants are bacteria, certain fungal partners are also very important for plants, as discussed further in Chapters 2 and 6.

Healthy animals and plants also harbour a diverse community of viruses, many of which have yet to be identified. In humans and many other hosts, most of the resident viruses are phage, i.e. viruses that attack bacteria, not eukaryotic cells (see box: **Microorganisms**). The phage may play a role in shaping the composition of bacteria in the microbiome (Chapter 2), offering the potential of phage therapies to treat some diseases (Chapter 7).

Why microorganisms promote the health of animals and plants

The previous section provided some examples of microorganisms that contribute directly to the wellbeing of animals and plants, by providing nutrients and by protecting them against natural enemies. There is, however, a further way in which microorganisms can be beneficial to their host. This is by influencing the function of hormones and other chemical signals that coordinate the activities of host cells and regulate key functions, including growth and development, immunity, and nutrient allocation. Evidence that the complex network of host signalling operates correctly only in the presence of microorganisms comes from research on laboratory mice raised under sterile conditions, so that they are perfectly free of any microorganisms. These mice are called 'germ-free', and they have many physiological defects. For example, the blood capillaries in their intestines fail to form properly, limiting the transfer of digested food from the gut to the bloodstream and the rest of the body, and the final stages in development of the immune system are blocked, leaving these mice with immature and incomplete defence against pathogens. When the germ-free mice are administered bacteria from other mice, most of these defects are corrected within two weeks.

Why are microorganisms needed for proper signalling between the cells and organs of mice and other animals? The likely answer is that the ancestors of animals were intimately associated with microorganisms, and the signalling networks that coordinate the function of the different animal cells evolved in the context of long-established interactions with microorganisms. In other words, we should not consider the microbiome as an add-on to the established form and function of an animal. The reality is that our ancestors were multi-organismal (i.e. living in intimate associations with bacteria and other microorganisms) long before they were multicellular. The same argument applies to the role of microorganisms in plant growth and health.

These considerations raise two questions that we will explore in the remainder of this chapter. First, why has it taken so long for scientists to appreciate the significance of the microbiome for health and disease in humans, other animals, and plants? Second, how is it possible that the microbiome, which is so important for good health, can also contribute to disease?

Microbiomes: hidden in plain sight

Our microbial partners were very evident to the first microscopists. Notably, Antonie van Leeuwenhoek, the 18th-century Dutch inventor of single-lens microscopes, reported microorganisms (which he called 'animalcules' or small animals) in human saliva. These and other observations by early microscopists were pursued by few investigators in the following years. This neglect can be attributed to two important scientific developments in the 19th and 20th centuries. The first was that, as evidence for the germ theory of disease accumulated, researchers focused on pathogenic microorganisms, and good health became synonymous with good hygiene and the absence of microbes. In parallel, early mathematical models in the developing discipline of ecology identified antagonistic interactions, especially competition and predation, as the drivers of population and community processes. Although associations of plants or animals with beneficial microorganisms were well known, mainstream biologists treated them as 'curiosities of nature' of no general significance.

These conceptual misunderstandings were exacerbated by a technical problem encountered by researchers attempting to study microbes associated with animals and plants. Microbiologists had developed sophisticated methods to isolate bacteria and eukaryotic microorganisms and then to study them as single-species cultures. The problem was that most microorganisms cannot grow under these conditions. As a result, microbiologists knew a great deal about the few species that could be cultured readily, such as the bacterium *Escherichia coli* and the baker's yeast *Saccharomyces cerevisiae*, but virtually nothing about >99 per cent of the microorganisms that were intractable to cultivation.

The breakthrough that led to the emergence of the discipline of microbiome science was technological. In the opening decade of the 21st century, automated methods for high throughput sequencing of DNA were developed, enabling microbiologists to sequence the DNA in any sample and, from this, identify members of microbial communities without culturing them in the laboratory (see box: **Molecular methods for studying microorganisms**). Cultivation-independent identification of microorganisms quickly became routine. Standard protocols are now available to identify the microorganisms in the leaf of a plant, a medical biopsy, the mucus released from a coral, a few grains of soil, and so on, and to infer their functional capabilities.

Molecular methods for studying microorganisms

The genetic material, specifically the sequence of DNA, is more similar in closely related organisms than in distantly related organisms, enabling us to display the relationships among many organisms as a phylogenetic tree, analogous to a family tree. The genes coding RNA molecules localized to ribosomes (the structures that mediate protein synthesis) are present in all organisms, and they are widely used to identify microorganisms in natural environments. The sequence of other genes is also used for microbial identification, as well as to provide information on functional traits (e.g. capacity to utilize a particular sugar or to synthesize vitamins) in microbial communities. Microbial ecologists use two complementary sequencing strategies. One is *PCR amplicon sequencing*, which involves the amplification of a specific DNA sequence (e.g. of a ribosomal RNA gene) in all the organisms in a sample by a technique known as Polymerase Chain Reaction or PCR, followed by sequencing and comparison to sequences already assigned to many different species in publicly available databases. Bacteria are routinely identified from the sequence of their gene coding 16S ribosomal RNA (16S rRNA; originally named from its sedimentation rate, 16 Svedbergs (S),

(*continued*)

during isolation by centrifugation). The second approach is *shotgun metagenomic sequencing*. The DNA is fragmented randomly into many pieces (analogous to the fragmentation of a shotgun shell when a gun is fired) and then sequenced, providing information on the presence and abundance of genes coding different functions. Readers interested in additional information are referred to the Further reading section for texts that explain these molecular technologies and their application to microbiomes, as well as some of their limitations (e.g. routine molecular methods fail to discriminate between living and dead microbial cells and can yield biased estimates of the abundance of different taxa).

The greater part of microbiome research conducted today is biomedical. This arises from the recognition that microbiomes, despite their beneficial effects on the host, can contribute to chronic disease. Let us now turn to address the relationship between microbiomes and disease, and how these interactions can be studied.

Relationships with microbiomes can turn sour

Although resident microbiomes generally promote the good health of their animal and plant hosts, these associations can 'go wrong' for the host. The overwhelming evidence that the microbiome can be deleterious contradicts the traditional but erroneous view that host-associated microbes can be assigned to two alternative categories: the pathogens, which cause disease; and the so-called 'friendly microbes' which promote good health. The reality is more complicated. Some individual microbial species can be beneficial or deleterious, depending on the age, genetic make-up, and lifestyle of the host. This is illustrated by a remarkable bacterium, *Helicobacter pylori*, which thrives in the acidic conditions of the human stomach. It generally infects humans early in life, with evidence that it promotes healthy development of the immune system in children. In older people, however, *H. pylori* can become deleterious, causing stomach ulcers and predisposing to stomach cancer. The incidence of *H. pylori* in many human populations declined dramatically in the 20th century, probably due to the widespread treatment of childhood illnesses with antibiotics, and it has been argued that this decline may be linked to the increased incidence of allergic diseases in children, as well as the reduced incidence of stomach ulcers in middle age.

Further complexity arises because many aspects of the biology of one member of a microbial community are strongly influenced by the activities of nearby microbes. For example, many bacteria in the human colon degrade dietary fibre, generating a diversity of small molecules known as metabolites. One of these metabolites, butyric acid, is very important for host health because it reduces gut inflammation and promotes gut barrier function. The amount of butyric acid produced in the colon is determined not only by the abundance of bacteria capable of producing this metabolite, but also by a web of interactions between the bacteria producing butyric acid and other microorganisms that lack this capability. There is evidence that perturbation of the microbial interactions that support butyrate production may contribute to inflammatory bowel disease.

Dysbiosis, disease, and interpreting correlations

The recognition that members of the microbiome can be deleterious to the host underpins much current interest in microbiomes, especially in relation to human health. The concept of dysbiosis provides a useful framework to explore this topic further (Figure 2). A healthy host is considered to be in balance with its microbiome, a state that is known as homeostatic. If this

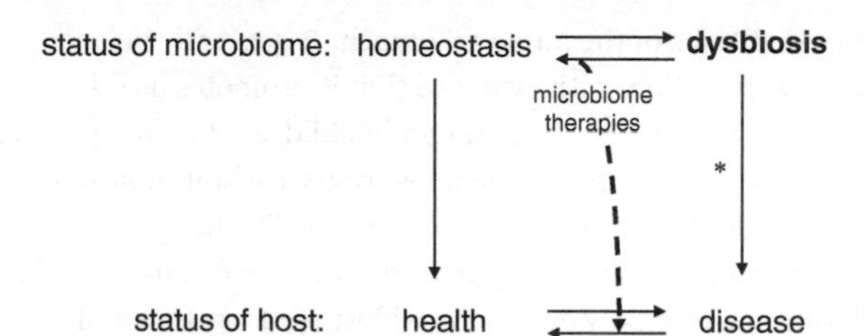

2. Dysbiosis of the microbiome and host disease. A dysbiotic microbiome is one that can cause or promote disease (asterisk). The goal of microbiome therapeutics is to modify the microbiome to the homeostatic state and, thereby, to ameliorate disease symptoms and ideally to cure disease (broken arrow).

condition is perturbed, for example by a major change in the lifestyle of the host or exposure to antimicrobial chemicals, the microbiome can change to an alternative state, known as dysbiotic, that is defined as a microbiome that can trigger or exacerbate disease (Figure 2). It is argued that microbial therapies which shift the community back to a homeostatic state should ameliorate or even cure the disease. These therapies include administration of the desired microorganisms, for example in probiotic foods, or eating foods enriched in specific types of fibre that promote the desired microorganisms.

Indications that the composition of microbiomes might influence host health come from many reports that the microbiomes of people with certain diseases differ from the microbiomes in healthy people. These differences are especially pronounced for the gut microbiome in people with type 2 diabetes, inflammatory bowel disease, autistic spectrum disorder, some auto-immune diseases, and colo-rectal cancer. However, the relationship between microbiome composition and disease is just a correlation, and it is open to two alternative explanations. Change in the microbiome to a dysbiotic state may trigger or exacerbate disease; or, alternatively, the disease may cause changes to the microbiome that have no effect on patient health.

Why microbiome research is complicated

The contribution of the microbiome to human health and disease is a major preoccupation of microbiome researchers, and rightly so. Microbial therapies have the potential to resolve many chronic and life-threatening diseases. Nevertheless, microbiome research is complicated. Here, I outline the key strategies that are adopted, and consider why interpreting results can be very difficult.

Let us start where we left off in the previous section: with a correlation between a human disease and the composition of the microbiome. The most straightforward way to establish whether the microbiome causes the disease is by experiment: to administer samples of the microbiome of a patient to a healthy host and investigate whether the recipient host develops disease. Of course, it would be both unethical and illegal to conduct this experiment on humans. Instead, many researchers use laboratory mice. Germ-free mice (i.e. mice that are raised under microbe-free conditions) are colonized with a microbiome sample from people who are healthy or have a disease. Results showing the development of the disease in mice with the microbiomes from unhealthy people, but not in mice bearing microorganisms from healthy people, provide strong evidence that microbiome composition can cause disease.

Laboratory mice colonized with microorganisms from humans are referred to as 'humanized mice'. Although it is straightforward, in principle, to interpret results of studies on humanized mice, there are several problems. It is important to understand some of the complexities, so that we do not come to over-simplistic conclusions about the effects of the microbiome on human health.

One problem in interpreting studies on humanized mice is that not all the microorganisms in the human microbiome thrive in the mouse gut. If the microorganisms that cause disease fail to transfer, a mouse study will not detect any microbial contribution to disease; this is a false negative result. A related issue is that some microorganisms may interact differently with humans and mice, causing disease only in humans (another false negative) or only in mice (a false positive).

A second difficulty relates to the choice of microbiome samples from healthy people to use as the control in mouse studies. The composition of the gut microbiome varies widely among healthy people. Some studies seek to match control samples and patient samples, for example by sex, age, and lifestyle, and this can reduce spurious differences. To illustrate the complexity, evidence from one study indicates that much of the difference between the microbiome of people with type 2 diabetes (T2D) and healthy controls could be explained by the tendency of the T2D patients to abstain from drinking alcohol for health reasons. The researchers found that whether and how much alcohol a person drinks was more important than disease status in explaining the microbiome differences between the T2D and control samples. Only time will tell whether these findings are general or specific to the cohort of people used in this study.

There is a further issue that besets almost all studies of the human gut microbiome: that human gut microbiome samples are usually harvested from faeces. Some microorganisms are shed in faeces more copiously than others, and the species that are shed at high rates will appear (erroneously) to be more abundant in the gut than those that are shed at low rates. Compounding this issue, the faecal microbiome is dominated by microorganisms in the colon, and the less abundant communities of microorganisms in other regions of the gut, such as the small intestine, are often not detected or recognized, even though they can play a critical role in health and disease.

How can these difficulties be overcome? The solution is to identify multiple, different ways to investigate a specific question, and to draw conclusions only where the answers obtained by the different approaches are the same or at least point in the same direction. Progress in microbiome research is slow and painstaking, requiring multiple lines of evidence. Recognizing this protects us from believing claims that microbiomes can explain everything—or nothing.

In the following chapters of this book, I consider how the microbiome affects the health of humans and other animals, focusing on nutrition and metabolic health (Chapter 3), behaviour and mental health (Chapter 4), and immunity and infectious disease (Chapter 5). In Chapter 6, I turn to the role of microbes in plant growth, including crop production, and Chapter 7 provides an overview of the ways we can harness microbiomes for improved health and the public good. But, first, let us consider how hosts acquire their complement of microorganisms (Chapter 2).

Chapter 2
How to get and keep a microbiome

Most animals start life as an egg, and plants start as a seed or spore. For the many species that harbour few or no microorganisms during embryonic development, the transition to independent life involves the abrupt exposure to myriads of microorganisms, and for these microorganisms, hatchling or newborn animals and seedling plants are a brand-new habitat to exploit. The challenge for every new host is how to facilitate colonization by beneficial microorganisms, while deterring pathogens. Some hosts (but not humans), however, bypass the hazards of acquiring a suitable microbiome because their microbes are inherited faithfully from mother to offspring over countless generations. These inherited microbes are precious heirlooms.

This chapter explores the many ways in which animals and plants acquire microorganisms from the external environment and from other hosts, including the mother. We will begin with humans, then consider other animals, and finally plants.

Where the microbiome of human babies comes from

Birth is the greatest change in our lives. This event involves the dramatic transition from a dark, warm place with food and oxygen delivered directly to the bloodstream from the maternal placenta to an independent life requiring continuous breathing for oxygen supply and regular feeding for nourishment. Birth also marks the transition from an essentially microbe-free existence to a world that is teeming with microorganisms. The gut and skin of a newborn baby are colonized rapidly by microorganisms.

The first microorganisms associated with the skin of a baby are bacteria, especially *Lactobacillus* species, acquired from the mother's birth canal (vagina). Other bacteria are introduced by skin–skin contact with the mother and other carers; these include the dominant members of the adult skin microbiome (Table 1). Over the first year of life, the composition of the infant's skin microbiome stabilizes and comes to vary by location in a similar way to an adult. For example, moist locations, especially the arm pit, support *Corynebacterium* and *Staphylococcus* species, while dry skin, as on the hand, has a more diverse microbiome. However, the skin of children bears a greater diversity of microorganisms, especially bacteria and fungi, than that of adults,

Table 1. Some abundant microorganisms in the human microbiome

Location	Dominant microorganisms
skin	*Staphylococcus epidermidis* (Bacteria of the phylum Firmicutes), *Cutibacterium*, and *Corynebacterium* species (Bacteria of the phylum Actinobacteria) and *Malassezia* species (Fungi). (Compared to adults, the skin microbiome of children includes a greater abundance of Bacteroidetes and Proteobacteria (phyla of Bacteria) and a wider diversity of fungi.)
gut	Bacteria of the phylum Firmicutes (e.g. species of *Clostridium* and *Roseburia*, *Faecalibacterium prausnitzii*) and the phylum Bacteroidetes (e.g. species of *Bacteroides* and *Prevotella*).
vagina	*Lactobacillus* species (Bacteria of the phylum Firmicutes) in many women.

and it is only at puberty that the skin microbiome stabilizes to the adult composition.

As for the skin microbiome, the first gut microbiome of a baby can be dominated by *Lactobacillus* species acquired as the baby passes through the birth canal, and these bacteria are readily detectable in the meconium (the first faeces of a newborn). These bacteria do not persist and the gut microbiome of the infant rapidly comes to comprise a small subset of the bacteria in the adult gut. The exact composition of these first bacterial communities varies widely among individual infants, but *Bifidobacterium* and *Bacteroides* species tend to be abundant. The next change is at weaning, when there is a dramatic increase in diversity of the microbiome. At this time, bacteria that can utilize dietary fibre, degrade plant secondary compounds (chemicals that occur naturally in plants but are not nutritious to animals), and synthesize B vitamins become abundant. The principal source of these microorganisms is the mother, other family members, and carers. Over the following months, the composition of the microbiome is very variable, both in a single baby over time and among babies of the same age. This variability gradually declines, and the microbiome in most 3-year-old children is broadly stable and similar to the microbiome of adults, i.e. comprises 500–1,000 species of bacteria and is dominated by two major groups, the Firmicutes and the Bacteroidetes (Table 1).

There is one further issue relating to microbiome acquisition in humans. It has been claimed that the foetus is exposed to bacteria that colonize the placenta, umbilical cord, and amniotic fluid. However, the small amounts of bacterial DNA detected in samples of these tissues may be contaminants, including fragments of dead microbial cells that may cross the placenta. Although this issue is not fully resolved, the healthy foetus in humans is most probably microbe-free, as also applies to other mammals. For example, the standard protocols to generate germ-free laboratory mice by Caesarian section under aseptic conditions would not be possible if the mouse foetus were colonized with microorganisms *in utero*.

Two practices surrounding the birth and early care of human babies have a major impact on the microbiome: the mode of delivery and breastfeeding.

Mode of delivery and the microbiome

In medicine, Caesarean section (C-section) is a critically important intervention for the health of the mother or baby in an estimated 15 per cent of deliveries. Nevertheless, there is evidence that C-section delivery can have long-term health implications for the child, through to adulthood, including a greater propensity for immunological disorders, such as asthma, inflammatory bowel disease, and type 1 diabetes. Heightened risk of ill-health is expected for C-section deliveries because of a greater incidence of pre-existing medical conditions in this cohort of mothers and babies than among vaginal deliveries. However, when this confounding factor is accounted for by statistical techniques, a significant difference in health outcome between C-section and vaginal delivery remains.

Much current research is focused on whether and how deficiencies in the gut microbiome of C-section-delivered infants may contribute to ill-health. The first microbiome of the skin and gut of babies delivered by C-section is dominated by bacteria derived from adult human skin. The difference between the microbiome of babies with C-section and vaginal delivery gradually declines and, generally, no reliable microbiome differences can be detected between the two groups by the first birthday. Despite these encouraging findings, there are indications that the perturbed gut microbiome of babies delivered by C-section in the first months of life can impair the maturation of the immune system. In principle, these difficulties can be overcome by administering health-promoting microorganisms to C-section babies, and this issue is the subject of intensive research.

Human milk and the gut microbiome

The gut microbiome differs between breastfed babies and babies fed on formula milk (which is derived from cow's milk). There are much higher levels of certain bacteria, especially *Bifidobacterium* species, and lower overall diversity of bacteria in the gut microbiome of breastfed babies. It is widely recognized that certain *Bifidobacterium* species support infant health by protecting against pathogens and regulating immune functions, and this most likely contributes to the low incidence of allergic diseases in children that were breastfed as babies.

The distinctive composition of the gut microbiome in breastfed babies can be attributed to the remarkable composition of the sugars in human milk. The dominant sugar in the milk of all mammals is lactose, which is made up of two sugar units, glucose and galactose. Lactose is easily digested by baby mammals, and a major source of energy for growth. Milk also contains small amounts of other sugars, including oligosaccharides, which comprise three to six sugar units. The oligosaccharide content of human milk is very high (5–20 g per litre), and includes three additional components (fucose, acetylglucosamine, and a sialic acid) which, along with glucose and galactose, are linked together in many different ways. More than 100 different kinds of human milk oligosaccharides (HMOs) have been described.

HMOs are important because, unlike lactose, they cannot be digested by the baby. They are passed without modification into the colon, where they are utilized by bacteria, especially *Bifidobacterium*. The different HMOs support the growth of different *Bifidobacterium* species and strains, and some HMOs can only be broken down by several species working together.

Because of these HMOs, human milk is a natural prebiotic food, meaning that it feeds beneficial microorganisms in the gut. Interestingly, the HMO composition of human milk varies among different mothers, and this may contribute to the variation in the gut microbiome composition among infants. There is no evidence that the differences in HMO profiles in the milk of different mothers influence the health of their babies.

As well as promoting beneficial bacteria, HMOs can suppress some harmful bacteria. For example, they inhibit the growth of group B *Streptococcus*, which can cause sepsis and meningitis in newborn babies. HMOs also prevent the attachment of diarrhoea-causing bacteria, such as *Campylobacter*, to the gut wall of babies, thereby protecting infants against diarrhoeal illnesses.

The high levels and complexity of oligosaccharides of human milk is most unusual, compared to the milk of other mammals. The milk of some other primates is enriched in oligosaccharides, but less so than in humans. The oligosaccharide content of cow's milk is 100–1,000 times lower than in human milk, and this is a major factor contributing to the low abundance of *Bifidobacterium* in babies fed on formula milk.

A microbiome as unique as your fingerprint

The gut microbiome in each person is unique. The reason is that, although the human population is very large, approaching 8 billion, the diversity of microorganisms that can inhabit our guts is much, much greater. Each person harbours up to 1,000 bacterial species, as well as various eukaryotic microorganisms, especially fungi, and many of these species are represented by multiple strains. The microbial community in one individual is a tiny fraction of the total number of compatible species and strains, and so the chance of two people bearing the same set of microorganisms is infinitesimally small.

It is often said that an individual's microbiome is as unique as their fingerprint. This statement is correct, but incomplete. A fingerprint is fixed, but the microbiome can change over time. Much research has been conducted on the gut microbiome, by analysing the microorganisms in faecal samples of individual volunteers taken regularly over many weeks or months. (The methods are described in the box **Molecular methods for studying microorganisms** in Chapter 1.) The composition of the microbiome varies somewhat, but not very much, from one sample date to the next. In other words, the gut microbiome of an individual human exhibits consistency, but not constancy.

The details of the time course studies of the gut microbiome are illuminating. Some microorganisms are detected in most or all faecal samples, and these are interpreted as permanent residents. Others come and go; they persist for up to several weeks, then disappear, and may return at a later date. These are microorganisms that can colonize the gut but cannot sustain long-term populations. The last group are detected in just one or two consecutive samples, and these are transients that pass through the gut with food.

The beneficial effects of the gut microbiome on human health are generally mediated by the permanent residents, rather than by the microorganisms that are only intermittently present or are transient. Interestingly, the composition of the permanent residents varies widely among different individuals. In other words, there is no core set of permanent microbial residents in all healthy people. Our genetic make-up, health status, and lifestyle can influence the suitability as a habitat for different strains and species of microorganisms, but at least a part of the among-individual variation may be attributable to chance: that the permanent microorganisms are those that happened to colonize first. Very long-term data are not available, but it is entirely possible that the permanent residents identified in studies conducted over one to two years in adults may include the descendants of microorganisms acquired in early childhood.

What you eat and the company you keep

Many researchers are investigating the factors that promote and resist change in the gut microbiome. This knowledge is critically important for designing strategies to shift the microbiome composition in an unwell person from a dysbiotic state towards a composition that promotes health (see Figure 2). Conversely, an understanding of the processes that resist change in the microbiome can be harnessed to design protocols that protect beneficial microbes in people administered antibiotics or chemotherapy.

Various aspects of a person's lifestyle may influence the composition of their gut microbiome, and frequently they are interlinked. Among these, two major contributing factors have been identified in many studies: diet and social contacts.

Individuals with a diet rich in plant fibre favour a species-rich gut microbiome that is generally associated with good metabolic and immunological health, while diets that include ultra-processed foods with high levels of sugars and fat promote a low diversity microbiome, sometimes dominated by species correlated with obesity, type 2 diabetes, and cardiovascular disease. Generally, a major shift in diet alters the microbiome, and the change can be very rapid. For example, one study investigated how the gut microbiome of 10 volunteers responded to two test diets, each taken for just five days: a plant-based diet of grains, vegetables, and fruits, and an animal-based diet of meat, eggs, and cheese. For all 10 individuals, the composition of the microbiome shifted in much the same way over the five days on these diets, and then returned to the pre-experiment composition after the individuals resumed their usual diet. One of the volunteers was a lifelong vegetarian. Before the experiment, the composition of his microbiome differed from the other nine individuals. It shifted strongly during the five days on the animal-based diet—and then reverted to normal within six days of return to his standard vegetarian diet.

There is also evidence that, as for pathogens, beneficial microorganisms can be transmitted by social interactions, including direct contact between individuals and via shared living space. Whether we consider the microbiome of the skin, the mouth, or the gut, members of one household are more similar to each other than to other households. In many societies, the social networks are large and complex, and it would be a daunting task to investigate the transmission of microorganisms through social contacts beyond individual households. These difficulties have been circumvented in a detailed study of small rural communities on the Fiji Islands, where the lifestyle (including diet) is relatively uniform and the social networks are quite simple. Analyses of the microorganisms in oral samples and faecal samples from members of these communities demonstrated both high within-household similarities and greater similarities between individuals from different households with strong social contacts (near neighbours or work companions) than weaker contacts, for example between members of different villages.

Much has yet to be learned about the transmission of beneficial microorganisms via social contacts. Questions that are being investigated include: how are they transmitted? How frequently? Are some microbial species transmitted more readily than others? Are some people 'superspreaders' of beneficial microbes? Resolving these questions is important because microbiome composition can influence our health. The company we keep can strongly affect our health and happiness—for microbiological reasons, as well as social reasons.

Kill the winner

We have seen that a varied and fibre-rich diet promotes a healthy and diverse gut microbiome. Other factors are also important, and these include interactions among members of the microbiome. Of particular interest are phage, viruses that attack only bacteria (see box: **Microorganisms** in Chapter 1). Phage viruses are very abundant in the human gut microbiome, with 1–10 billion phage particles shed per gram of faeces from a healthy adult, and most phage are specific to a few strains or species of bacteria.

There is growing evidence that phage may play an important role in controlling the numbers of bacteria in the gut microbiome. It is reasoned that, as a bacterial species becomes more abundant, the phage that specialize on those bacteria are more likely to contact a susceptible bacterial cell, and then kill it, releasing millions of infective progeny phage particles, each of which can infect another bacterial cell of the same species. This process suppresses the most abundant bacteria; it is often referred to as 'kill the winner'. In this way, phage can prevent overgrowth of individual bacterial species (provided phage specific to that species are present), and this can lead to an overall increase in the diversity of microbial species in the microbiome, which is generally beneficial for host health. Although the role of phage in the gut microbiome of humans and laboratory mice has been studied most intensively, these interactions are also likely to be important for the skin microbiome and the microbiomes of other animals and plants.

How non-human animals acquire their microbiomes

Broadly speaking, there is nothing exceptional about the routes by which humans acquire and retain their microbiomes. Other newborn mammals gain their first microbiome from contact with their mother; in many species, transmission is facilitated when the mother repeatedly licks her newborn. Social contacts can also promote the exchange of microorganisms throughout life. This is illustrated by a detailed study of wild baboons in Amboseli Park, Kenya. Baboons live in large social groups, and group members interact repeatedly by grooming each other. Although many aspects of baboon life, including diet, are uniform for the entire group, each grooming relationship tends to be restricted to a few individuals. The gut microbiome is much more similar in pairs of baboons that groom each other extensively than in baboons that rarely or never groom each other, strongly suggesting that the repeated physical contact of grooming promotes the transfer of microorganisms between individuals. Social contact is also implicated in the exchange of microbes in other animals. For example, microbiome experiments conducted on laboratory mice have to be designed very carefully to take into account that the microbiome in mice from the same cage tends to be more similar than in mice from different cages; this pattern is found even when the mice in multiple cages are genetically identical and fed on the same foods under strictly standardized conditions.

Social interactions can also be crucial for the acquisition of microbiomes in animals other than mammals. One of the most complex societies in the animal kingdom is found in honey bees. Within the bee hive, a single queen lays many eggs. Most of the eggs develop into female worker bees that collect food (flower nectar and pollen), maintain the hive, and care for the developing young, which are called larvae. The gut of adult bees bears bacteria that protect the insects against pathogens and may provide nutrients. However, the bee larvae are essentially microbe-free. In the first few days after reaching adulthood, the worker honey bee feeds on bacteria-rich droplets of hindgut fluid released from the anus of older worker bees. The bacteria pass unharmed through the stomach and intestine to the hindgut, where they adhere to the gut wall and proliferate to form a protective layer.

Many animals are not social and do not care for their young. For these animals, there is limited opportunity to acquire beneficial microorganisms via contact with other members of the same species. In many frogs, fish, insects, and worms, the microbiome is dominated by the subset of the microorganisms in the external environment that can tolerate the conditions and use the resources in the animal gut or on the animal's surface. In some animals, just a single species is acquired. For example, the luminescent bacteria that occupy the light organ of the bob-tailed squid comprise members of one species, *Vibrio fischeri*. Among the myriads of different bacteria in the water column of the squid habitat in the shallow seas of Hawaii, only this species can navigate through the light organ pore on the squid surface, survive the antimicrobial defences (including bleach and reactive oxygen species) in the narrow light organ canal, and then proliferate in the light organ.

To depend on environmental sources of the microbiome can be risky, especially where a particular microbial species is absolutely required by the animal host and is not universally present in the environment. Unsurprisingly, highly stereotyped and sophisticated mechanisms to transfer beneficial microorganisms from mother to offspring have evolved in various animals. For example, nothing is left to chance in cockroaches, which derive essential nutrients from just one bacterium, *Blattabacterium cuenoti*. This bacterium is unknown in the external environment, and cockroaches acquire it from their mother. A small inoculum of *B. cuenoti* is transferred from the symbiotic organ to the ovary, and then into each egg, so that the relationship is set up and established even before the cockroach offspring hatch from the egg. In fact, there is strong evidence that *B. cuenoti* has been transmitted from cockroach mother to offspring in this way for more than 140 million years. This habit probably arose in the Jurassic period, at a similar time to the origin of birds and probably before the first flowering plants evolved, and it has persisted through every generation of cockroaches since then. The transmission of specific bacteria via the ovary has evolved independently in other insect groups, including some ants, the tsetse flies, blood-feeding lice, and various plant lice, such as aphids and whiteflies. As for the cockroaches, many of these relationships are very ancient. Humans and other vertebrates do not acquire microbes in this way.

Rooted to the spot: how plants acquire microbes

Healthy plants harbour large and diverse microbial communities, the composition of which differs between three broad locations: the internal tissues of the plant (the endosphere microbiome); the surface of the shoots, including the stems, leaves, and flowers (the phyllosphere microbiome); and soil immediately adjacent to the surface of the roots (known as the rhizosphere microbiome). Although some microbial species are found in two or all three of these locations, the overall microbial diversity is greatest in the rhizosphere microbiome, much lower in the shoot, and often limited to just a few species in the endosphere.

The rhizosphere microbiome is derived from the exceptionally diverse microbial communities detected in bulk soil (i.e. soil at a distance from plant roots). A gram of soil is estimated to bear up to 5 billion bacterial cells, often representing more than 10,000 species, and the composition of the microbial communities can differ widely between different soil samples. Two sets of processes determine the composition of the rhizosphere microbiome associated with any one plant and explain why the microbiomes vary between different plants. The first is chance: only a tiny fraction of the microorganisms living in the soil make contact with a specific plant. The second set of processes comes under the umbrella term plant factors. Plants can exert a degree of control over the microorganisms in the rhizosphere, by a mix of inducement and deterrence—a strategy of 'carrot and stick'.

Because plants are literally rooted to the spot, they are preferentially colonized by microorganisms which come to them. Plant roots in undisturbed soils are colonized principally by motile bacteria that swim through the fine film of water surrounding soil particles, and then proliferate on the root surface and nearby soil. Some bacteria (notably Actinobacteria) and many fungi form multicellular filaments, known as hyphae, and these hyphae grow towards plant roots. The plant root influences the direction of microbial movement and growth by the continuous release of exudates that contain both inducements, including nutritious sugars and amino acids, and deterrents, especially selective antimicrobial compounds. Plants use up to 20 per cent of the carbon that they capture by photosynthesis to make sugars that are released in root exudates. The allocation of so much carbon to exudates, instead of to growth, is a sure sign that root exudates are important to the plant.

Much has been learned about how root exudates can shape the composition of the rhizosphere microbiome from detailed studies on a species of cress, *Arabidopsis thaliana*. The root exudates of *Arabidopsis* include up to 50 different kinds of complex organic compounds called triterpenes. Any one of these compounds promotes the growth of some soil bacteria but suppresses the growth of others. Although the exact set of bacteria favoured can depend on the precise blend of triterpenes, beneficial species tend to be favoured, and pathogens suppressed by the released compounds. Laboratory experiments have used *Arabidopsis* that have been genetically modified to produce different triterpenes and then exposed to various defined sets of bacteria. The composition of the resultant rhizosphere microbiome associated with these *Arabidopsis* plants could be predicted from the growth pattern of the various bacteria incubated in pure culture with different triterpenes.

These laboratory experiments on *Arabidopsis* provide proof-of-principle that individual compounds and classes of compounds in root exudates can shape the rhizosphere microbiome. The relationship between plants and their rhizosphere microbiome in the natural world is much more complicated because root exudates contain many different classes of compounds (i.e. not only triterpenes) that can vary widely between different plants and with root age and environmental conditions. Potential microbial colonists of plant roots have to negotiate a complex and ever-changing array of plant chemical inducements and deterrents.

A harsh life on leaves

The stem or leaf of a plant (often known as the phyllosphere) is a place of hardship for microorganisms. It is not just that microbes on the surface of the aerial parts of plants are exposed to harsh conditions of sunlight, which includes UV radiation, and extremes of water availability and temperature. It is that the conditions can change dramatically over short time periods, from day to night, or with a passing cloud or rain shower. There should be no surprise that the phyllosphere microbiome is relatively small, up to 1,000 times smaller than the rhizosphere microbiome.

Where do the phyllosphere microbiomes come from? Some come from the soil, with evidence from laboratory studies on *Arabidopsis* that certain bacteria migrate from the rhizosphere to the emerging shoot of seedlings. Other members of the phyllosphere microbiome are derived from above-ground sources, including wind-blown aerosols and raindrops, as well as falling from the surface of insects and other animals visiting the plant. The patterns of transfer are complex. For example, more microorganisms are lost than gained by individual leaves during rain, but more microbes are deposited from insects moving around on wet leaves than on dry leaves.

Most of the microorganisms deposited onto leaves fail to survive, for lack of nutrients and water. Bacteria that are most likely to persist are pigmented, for protection against strong sunlight.

In addition, many phyllosphere bacteria release detergent-like substances, which make the plant surface more wettable, helping bacterial cells to move around and access nutrients. Exactly where on the leaf surface a microbial cell happens to land is also important. A leaf is complex terrain, with deep ruts at the junctions between adjacent plant cells, elevations along leaf veins, holes (the stomatal pores) for the movement of carbon dioxide and water vapour, and a variable number of hairs (trichomes), some of which secrete chemicals. The sites that are best protected from the elements and provide sufficient nutrients tend to be at the base of trichomes, adjacent to leaf veins and in the grooves between epidermal cells, where dense aggregations of microorganisms can accumulate. Most other regions of a leaf surface bear few or no microorganisms.

Gaining access to the internal tissues of plants

Over many decades, plant scientists have studied two groups of endosphere microbes (i.e. in the internal tissues of plants) that alter the morphology of plant roots. One is the rhizobia, a group of bacteria that inhabit swellings, known as nodules, on the roots of many legumes (plants of the pea family). Rhizobia fix atmospheric nitrogen into ammonia, enabling their plant host to thrive in nitrogen-poor soils. The other well-known endosphere inhabitants are mycorrhizal fungi, which were evident to early plant microscopists as hyphae ramifying between cells of the root cortex. One group of mycorrhizal fungi infects very many plant species and forms densely branching structures, known as arbuscules (they look like 'little trees') in the root tissue, facilitating nutrient exchange with the plant; these fungi are known as arbuscular mycorrhizal fungi (AM-fungi). A key function of the AM-fungi is to harvest soil phosphorus to support plant growth. In exchange, most mycorrhizal fungi obtain sugars from the plant.

Despite the morphological and functional differences between plant interactions with rhizobia and AM-fungi, these two types of

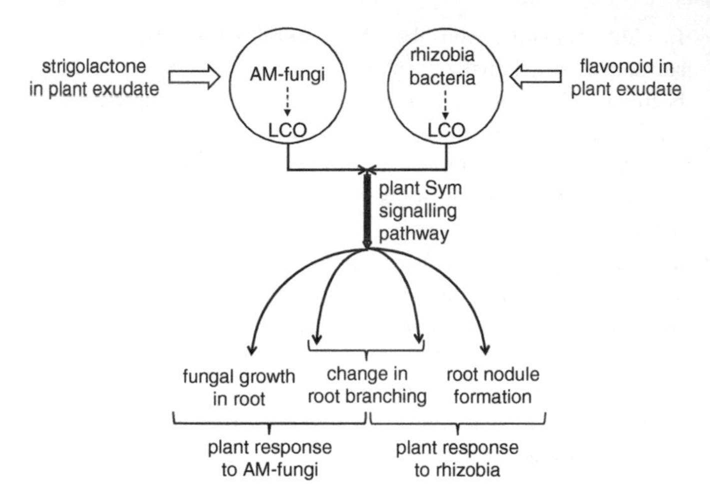

3. Common pathways for colonization of the plant root endosphere by AM-fungi and rhizobia. Suitable partners are recognized initially by the exchange of plant signals (strigolactone or flavonoid) and microbial signals (highly modified sugars, known as lipo-chito-oligosaccharides (LCOs) and, for AM-fungi, also chito-oligosaccharides (COs)). The common plant Sym signalling pathway orchestrates precisely timed changes in gene expression, cell biology, metabolism, and overall morphology of both the plant and microbial partner. The signals from AM-fungi and rhizobia are believed to activate the Sym pathway in subtly different ways, ensuring that plants capable of forming both types of association (e.g. many legumes) do not confuse the two types of partner.

microorganisms gain access to the plant root by remarkably similar molecular mechanisms (Figure 3). Both rhizobia and AM-fungi activate a common molecular signalling cascade in the root, instructing the root to make a nodule for the rhizobia and to change its morphology (especially root branching pattern) in response to the AM-fungus. This signalling is known as the Sym Pathway and it is activated by highly modified sugars released from the incoming rhizobia or AM-fungi. These sugars are referred to as LCOs (see caption to Figure 3). Going back one step further, the microbial partners release these sugars at high rates when they detect that they are close to the root of a suitable plant. The plant 'message of invitation' comprises specific compounds in their root exudate: flavonoids for rhizobia, and strigolactones for AM-fungi.

The plant Sym pathway is undoubtedly very ancient, probably derived from the signal exchange between microorganisms and the first plants that invaded land, 400 million years ago. Supporting this interpretation, there is both fossil and molecular evidence that the first land plants were associated with AM-fungi (see Chapter 6 for details). The ancient Sym pathway was subsequently recruited to facilitate the associations with rhizobia that evolved in legumes, less than 60 million years ago.

The chemical codes of microbial–plant associations

The conserved molecular conversation between the plant roots and both AM-fungi and rhizobia (Figure 3) is overlain by an elegant mechanism which controls precisely which of the very many possible rhizobia or AM-fungi is admitted to the internal tissues of the plant root. The detail of this mechanism is well understood for the rhizobia.

Many different species of rhizobia can be found in bulk soil and the rhizosphere of many plants, but each rhizobial species or strain is capable of inducing, and then inhabiting, root nodules in just a few species of leguminous plants. For example, *Sinorhizobium medicago* associates with alfalfa, while *Rhizobium leguminosarum* of one strain (*viciae*) colonizes pea plants. The 'right' plant and rhizobial partners swap chemical codes, the equivalent of movie spies who identify each other by reciting scripted questions and answers. The scripted 'phrases' for legumes and rhizobia are, first, a very specific plant flavonoid, to which only the 'right' rhizobial species or strain responds by releasing a specific LCO. If this is the 'right' LCO for the plant, a root hair responds by curling around the LCO-secreting bacteria. The enclosed bacteria divide repeatedly and are then taken into the root and transported to an internal region of the root known as the root cortex. While this is happening, molecular signals are sent along the Sym pathway to the root cortex, inducing plant cells to divide and to prepare the nodule for infection by the rhizobia. It takes approximately four weeks from initial contact for a functional nodule to develop.

The mycorrhizal fungi and rhizobia are not the only microbes in the endosphere of healthy plants. Many of the other members of the endosphere microbiome are acquired from the external environment (as for rhizobia and AM-fungi), but some microbes in the shoot are transmitted directly to the offspring plant via the seed. This has been demonstrated, for example, for some fungi in grasses and the bacteria *Burkholderia*, which form leaf nodules on certain plants, including *Ardisia* and *Styrax*. As the seed germinates and the seedling develops, the seed-borne microbes infect its shoot system. The molecular 'conversations' between plants and endosphere microorganisms other than AM-fungi and rhizobia appear to be diverse, some involving LCO-like sugars and others requiring chemically different signalling molecules.

The incentive to learn more about how certain microorganisms are admitted to the endosphere of plants is heightened by a growing appreciation that these associations can make valuable contributions to sustainable agriculture (see Chapter 6).

Chapter 3
Microbiomes, nutrition, and metabolic health

All organisms need adequate nutrients to grow and reproduce. The received wisdom from the discipline of nutritional science is that humans and other animals require a balanced diet, including essential nutrients which the animal cannot synthesize for itself. Advances in microbiome science are rewriting this traditional narrative—for some animals. Some resident microorganisms provide essential nutrients and contribute to the digestion of certain dietary constituents. The microbiome also functions to regulate many aspects of gut function and the wider nutritional health of the animal.

In this chapter, we will consider two broad issues: how the nutritional role of the microbiome can enable some animals to subsist on diets that professional nutritionists would define as grossly inadequate; and how disturbance to the microbiome can perturb nutritional health. In humans, as we shall see, there is increasing evidence that a dysbiotic microbiome can exacerbate stunting and impaired development of undernourished children, and contribute to nutritional ill-health in adults, increasing the risks of type 2 diabetes and cardiovascular disease.

Metabolism, nutrition, and 'the balanced diet'

The cell is the fundamental building block of living organisms. Every cell continuously synthesizes new materials, including proteins, lipids, and carbohydrates, for maintenance and growth, and generates energy to support these activities. These processes of building new cellular constituents and producing energy are mediated by a complex network of chemical reactions that is collectively known as metabolism (Figure 4).

The inputs to the total metabolism of an organism are nutrients. Many organisms need rather few types of nutrients. Plants can thrive when provided with water and carbon dioxide for photosynthesis, and inorganic sources of key elements, such as nitrogen, potassium, and phosphorus, and the much-studied bacterium *Escherichia coli* can proliferate indefinitely in medium containing an organic carbon source (e.g. a sugar) and just five inorganic compounds. Animals are different. All animals lack many metabolic capabilities, including the capacity to synthesize nine of the 20 amino acids (the nine essential amino acids) that make up proteins and many small organic molecules, including vitamins. Without each and every one of these essential nutrients, an animal cannot thrive and, ultimately, it will die.

A balanced diet for animals includes adequate sources of energy (carbohydrates and lipids), protein that provides all the essential

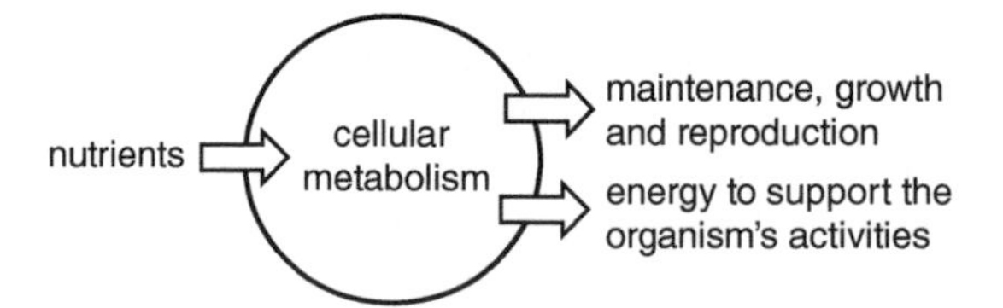

4. Metabolism and nutrients. Metabolism is the set of reactions in the cells of an organism that consume nutrient inputs to generate energy and the building blocks for maintenance, growth, and reproduction.

amino acids, vitamins, and various other organic micro-nutrients and minerals. The requirement for a balanced diet is a major drawback, limiting the range of diets and habitats that animals can exploit. How did this situation arise? The most likely explanation is that, in the distant past (probably >800 million years ago), the ancestors of all animals had access to a nutrient-rich diet: they probably fed mostly on bacterial cells. Over evolutionary time, they lost the capacity to make complex products, such as essential amino acids and the B vitamins, that were plentiful in their diet. The metabolic pathways to synthesize these nutrients are genetically and biochemically complex, and they have never evolved afresh in animals. The need for a balanced diet is the consequence of this inferior metabolic inheritance.

Nevertheless, some animals have escaped from the necessity for a balanced diet by associating with microorganisms that possess key metabolic capabilities. These interactions can be illustrated by research conducted on microbiomes in insects.

Resident microbes and the balanced diet

Consider the head louse *Pediculus capitis*, which lives in the human scalp, where it pierces the skin to feed on an exclusive diet of human blood. Unfortunately for us, the head louse suffers no ill-effects from this diet, even though blood is grossly deficient in B vitamins. The success of the head louse and the other species of lice feeding on the blood of various mammals can be attributed to an internal supply of B vitamins. Specifically, these insect parasites derive key B vitamins from bacteria that are housed in a specialized symbiotic organ, close to the gut. The symbiotic organ is white and readily visible against the bright red gut contents of a living louse that has recently fed. When the supply of B vitamins is eliminated, for example by surgical removal of the symbiotic organ or antibiotic treatment, the lice fail to develop and die. Other insects, including tsetse flies and bedbugs, as well as ticks, feed through the life cycle on vertebrate blood, and they also bear resident bacteria that provide B vitamins. Fleas and mosquitoes, however, lack these specialized associations because, although they feed on blood, they also use other food sources that contain an ample supply of B vitamins.

Blood is not the only nutritionally inadequate diet used by insects. Other food sources are grossly deficient in essential nutrients: plant sap used by plant lice, such as aphids and whiteflies; wood used by various beetles, including augur beetles and furniture beetles; and the diet of some generalist scavengers, such as cockroaches. As for the blood-feeding lice, these insects are associated with microorganisms in specialized symbiotic organs. These microorganisms are required for normal insect growth and reproduction, and they provide both essential amino acids and B vitamins.

Research on the microorganisms in symbiotic organs of insects has been important to our understanding of the nutritional and metabolic function of resident microorganisms. It has demonstrated that animals can successfully overcome their metabolic deficiencies, not by the evolution of novel metabolic pathways but by associating with specific microorganisms, usually bacteria, that possess the valuable metabolic capability. These insect associations are much simpler and more amenable to experimental study than many animal microbiomes. Generally, the symbiotic organ bears just one microbial species, in contrast to the hundreds of species associated, for example, with the gut or skin of humans or mice, and so the researcher does not have to work out which of many microbial species may be responsible for the metabolic function of interest. Studies are also simplified by the fact that, unlike many microbiomes which vary in composition between individuals, the symbiotic organ in every member of an insect species bears the same microbial species. This is because the microorganisms are transmitted faithfully from mother to offspring, usually via the egg. As considered in Chapter 2, some of these insect associations are very ancient, for example >100 million years old in cockroaches and many plant-sap-feeding insects, and bacteria with a specific nutritional function have been retained by the insect hosts as they evolve and diversify over these very long evolutionary timescales.

How relevant are these findings to the many animals that lack these specialized symbiotic organs, but have a diverse and abundant gut microbiome? Studies of gut microbiome in various insects have revealed that certain microorganisms can be nutritionally important. For example, the fruit fly *Drosophila melanogaster* thrives on diets that are deficient in certain B vitamins if the gut is colonized with *Lactobacillus* bacteria that can synthesize these micronutrients, but the *Drosophila* larvae fail to grow and develop on these diets if these critical bacteria are excluded. Similarly, the plant diet of turtle ants (*Cephalotes* species) is deficient in essential amino acids, and the ants derive a supplementary supply from a subset of the bacterial species in their gut microbiome.

Complementary evidence for the nutritional importance of the microbiome comes from a different line of research, the gut microbiome in herbivorous mammals, and I consider this topic next.

Plant food is not easy meat for animals

In principle, the most abundant source of food for terrestrial animals is cellulose, which is a polymer of the sugar glucose and the main constituent of plant cell walls. The big problem for an animal taking a mouthful of cellulose-rich grass, leaves, or wood is that the chemical degradation of the plant material to release the sugar is biochemically difficult, involving a complex battery of enzymes which function slowly. Anyone who maintains a garden composter has witnessed the long timescale of biological degradation of plant material.

Many animals have harnessed the capacity of microorganisms to degrade plant cell wall material. This is particularly evident in herbivorous mammals which, like all vertebrates, have minimal capacity to degrade cellulose. In most herbivores, including elephants, horses, rabbits, and voles, a portion of the hindgut is expanded as a large chamber that houses a complex community of microorganisms. (The precise anatomical location of the hindgut chamber is not uniform across species, e.g. in the colon of the horse, but the cecum of a rabbit.) The plant material delivered to this chamber has already been well chewed in the animal's mouth and then subjected to the animal's digestive enzymes in the stomach and small intestine. The hindgut community of microbes degrades the plant material to its constituent sugars and then uses the sugars as a carbon source for energy (Figure 5). Importantly, the hindgut chamber has minimal oxygen levels (like most regions of the mammalian gut), which means that the community of hindgut microbes display fermentation. The reactions are similar to brewing, where yeasts use sugars as a substrate to generate energy with the production mostly of alcohol; the differences are that the principal microorganisms in the mammalian hindgut are bacteria, and their main fermentation product is not alcohol but acetic acid, the chief constituent of vinegar. The acetic acid and other fermentation products are released from the bacteria in the hindgut, and absorbed across the gut wall into the animal's bloodstream. The cells of the animal, having access to oxygen, can oxidize the microbial fermentation products further to generate energy (Figure 5).

There is considerable current interest in the gut microbiome of animals that utilize cellulose-rich diets because the microbes in these animals can potentially be used to improve biofuel production from cellulose-rich biomass. Identifying suitable microbial species is a complex task because the microbial communities in the fermentation chambers of most mammals are very diverse, dominated by many Bacteria (especially members of the phyla Bacteroidetes, Firmicutes, Proteobacteria, and

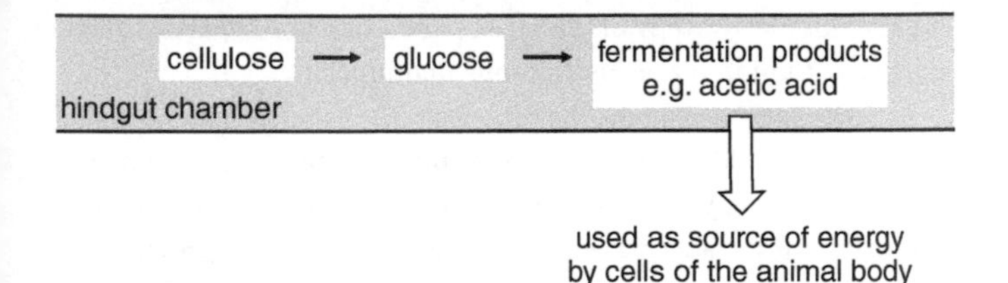

5. Degradation of cellulose in the hindgut of herbivorous mammals. The end products of microbial metabolism under oxygen-free conditions are known as fermentation products and they are transported via the bloodstream to cells of the animal body, where they are broken down with the aid of oxygen to generate energy. Other cell wall constituents (including hemicelluloses and pectins) are degraded in a similar way to cellulose.

Actinobacteria), but also including some Archaea, particularly methane-producing species, and various eukaryotic microorganisms.

Other animals that utilize plant material adopt different strategies. Some insects are associated with specific fungi that degrade plant material in the insect nest, not in the insect gut. The insects provide the fungus with plant material and feed on portions of the fungal growth. Examples include the leaf-cutting ants and the fungus-cultivating termites, both of which forage for plant material which is delivered to the fungal garden in the nest, and ambrosia beetles, which bore tunnels into trees and then line the tunnels with their wood-degrading fungal partners. Other animals are not dependent on microorganisms. For example, the wood-boring gribbleworms produce their own powerful cellulose-degrading enzymes, while many leaf-feeding caterpillars extract the easily digested nutrients and allow the cellulose in their food to pass straight through their gut. Intriguingly, the giant panda is unique among mammalian herbivores in feeding exclusively on plant foods but possessing a gut microbiota with little capacity to degrade cellulose. In explanation, the giant panda is a species of bear, and it has a very short evolutionary history as a herbivore.

The nutritional strategy of the giant panda has closer parallels to a caterpillar than to other herbivorous mammals.

Let's take stock. Microbiome studies indicate that microorganisms have enabled many animals to utilize diets that would otherwise be nutritionally inadequate. Resident microorganisms provide essential nutrients, for example in insects feeding on vertebrate blood and plant sap, and they can degrade dietary constituents, especially plant fibre, that are intractable to the digestive capabilities of many herbivorous animals. This leads us much closer to home: the significance of the gut microbiome to human nutrition.

Human nutrition and the gut microbiome

Most animals can be classified by their diet, for example as herbivore, omnivore, or carnivore, but there is no consensus about the appropriate category for the human diet. Humans are remarkably versatile in their eating habits, varying with geography and culture from predominantly meat-based to exclusively plant-based diets; but all human diets include foods that are processed, both physically, for example by crushing, and chemically, by heating or fermentation. The one point of agreement among professional nutritionists is that humans require a balanced diet that includes an adequate supply of all essential amino acids, and various other organic micro-nutrients, including vitamins. The gut microbiome does not appear to make a significant contribution to the human requirement for essential nutrients.

Nevertheless, microorganisms in the human colon mediate the degradation of dietary fibre, i.e. the components of plant-derived foods that are resistant to human digestive enzymes in the stomach and small intestine. As for the hindgut microbiome of mammalian herbivores (considered above), microorganisms in the human colon degrade the plant-derived material, generating a variety of fermentation products. These microbiome-derived compounds make a small contribution, probably 10 per cent, to the total energy requirements of a human, at least threefold less than in mammalian herbivores.

There is now excellent evidence that different microorganisms in the human hindgut degrade different types of plant fibre to fermentation products. As a result, a diet that includes a diversity of fibre-rich foods (fruits, vegetables, cereals, nuts, etc.) promotes a diverse microbiome. This, in turn, promotes human health because the fermentation products of the microorganisms have many positive effects on gut function, beyond their role as a source of calories. Microbial fermentation products have been implicated in promoting gut motility and the supply of blood capillaries to the gut. Some fermentation products also function as signalling molecules with many whole-body effects. For example, they regulate lipid storage and energy expenditure in various organs, including the liver and muscle, they dampen immune factors that cause inflammation in the gut and other organs, and they increase overall insulin sensitivity, thereby protecting against type 2 diabetes.

The evidence that the gut microbiome has a pervasive effect on the regulation of metabolic function has led to intensive research on the interaction between the microbiome and two aspects of malnutrition: childhood stunting; and metabolic syndrome, comprising obesity, insulin insensitivity, and cardiovascular disease. Let's consider these two disorders in turn.

Malnutrition in young children

More than 800 million people are undernourished. This shocking global statistic includes nearly a quarter of the world's children under 5 who are stunted by inadequate nutrition, with lifelong effects on their health. There is growing evidence that the gut microbiome in undernourished children is perturbed, exacerbating the effects of poor nutrition. As we noted earlier, a microbiome that is detrimental to health is described as dysbiotic.

A particularly severe form of undernutrition is kwashiorkor, caused by a protein-deficient diet. Children with kwashiorkor are stunted, emaciated, and have a swollen belly, and they tend to respond poorly when provided with a nutritious diet. The role of the gut microbiome in kwashiorkor is indicated by a study conducted on young children in Malawi. Faecal samples were collected from pairs of twins, only one of which displayed persistent symptoms of kwashiorkor, despite being given a nutritious food supplement. Although each pair of twins lived in the same household and ate the same food, their microbiomes were very different. Specifically, the microbiome of the children with kwashiorkor was remarkably stable and of immature composition, failing to display the normal developmental changes in healthy children (as described in Chapter 2).

To investigate whether the immature microbiome in children with kwashiorkor might contribute to the disease, samples of the microbiome from the healthy and affected twins were administered to germ-free laboratory mice. The mice were fed on a diet representative of the food eaten by the children. The results were very striking. The mice with the microbiome from the healthy twins maintained their weight, but those administered the microbiomes from the unhealthy twins lost, on average, 40 per cent of their weight. These experiments on laboratory mice provide the strongest evidence that kwashiorkor is not a simple nutritional disease caused by inadequate dietary protein, but also involves the establishment of a very stable but deleterious microbiome. The knowledge that these children are suffering from dysbiosis provides a basis for more effective, microbiome-based treatment of kwashiorkor.

For many undernourished children, the symptoms are more subtle than kwashiorkor. Of particular concern is a syndrome known as environmental enteropathy, which is very widespread in poor communities living in unhygienic conditions and is a major cause of childhood stunting, linked with increased susceptibility to disease, poor responsiveness to oral vaccines, and impaired intellectual development. The basis of this disorder is inflammation of the small intestine, the region of the gut between the stomach and the hindgut. In a landmark study conducted in the urban slum Mirpur in Dhaka, Bangladesh, the small intestine microbiome of 80 young children was sampled by endoscopy. The composition of the microbiome was determined, and it was discovered that the children with healthy growth had lower levels of 14 bacterial species than the children with stunted growth. These 14 bacteria were then administered to mice, which responded with inflammation of the small intestine and impaired growth, indicating that environmental enteropathy is a disorder of the microbiome. Based on this information, the researchers were able to develop a food supplement that, unlike standard nutritional supplements, supported a health-promoting microbiome of the small intestine and improved infant health. I return to this issue of microbiome therapy for improved childhood health in Chapter 7.

The microbiome and metabolic syndrome

In many countries, the most important source of nutritional ill-heath and premature death is not undernutrition, but overnutrition. The main culprit is the diet widely adopted as part of the industrial lifestyle: a diet dominated by energy-dense processed foods with little fibre and often depleted in protein and essential micro-nutrients. This diet, together with a sedentary lifestyle, has been linked to a dramatic rise over recent decades in the prevalence of a disorder known as metabolic syndrome. For example, data published by the American Medical Association in 2020 (covering up to 2016) shows that 35 per cent of adults in the USA, including 50 per cent of individuals aged over 60 years, are estimated to have metabolic syndrome. The incidence of metabolic syndrome is increasing worldwide, including in Europe, Latin America, South-East Asia, and especially the Western Pacific.

For the physician, the chief indicators of metabolic syndrome are high levels of blood glucose, unbalanced lipid content of the blood, specifically too much triglyceride and/or too little high-density lipoprotein, and high blood pressure. People with three or all four of these traits are diagnosed as having metabolic syndrome. They have increased risk of type 2 diabetes and cardiovascular disease. Most, but not all, people with metabolic syndrome are overweight or obese.

Metabolic syndrome is linked to perturbation of the gut microbiome. Specifically, poor diet and lack of exercise is argued to cause changes in the gut microbiome to a composition that exacerbates the negative effects of lifestyle on health, i.e. the microbiome becomes dysbiotic. Many studies have shown that the composition of the gut microbiome differs between lean and obese people. This effect is even apparent in comparisons between identical twins, one of which is obese and the other lean, indicating that this effect cannot be explained by differences in the genetic make-up of lean and obese people. When samples of the faecal microbiome from the twins were administered to mice, the mice with the microbiome from the obese individuals developed more fatty deposits. These experiments demonstrate how the microbiome can influence whether host metabolism is structured to favour fat deposition and obesity.

Unsurprisingly, there is much interest in identifying which of the 500–1,000 species of gut bacteria influence fat deposition in humans. A promising lead is that the microbiome of obese people tends to be depleted in bacteria of the family Christensenellaceae (in the phylum Firmicutes). In one study, the microbiome of an obese human donor with barely detectable levels of Christensenellaceae bacteria was introduced to germ-free mice.

As expected, fatty deposits accumulated in these mice. A second treatment in this study was mice administered the microbiome sample from the obese donor mixed with *Christensenella minuta*, isolated from a lean person; these mice were protected against excessive fat deposition. The basis of this effect is still unclear because the *C. minuta* supplement triggered complex changes in the abundance of many other microbial species. In other words, the beneficial effects of Christensenellaceae could be indirect, mediated by its effects on other microorganisms that influence energy metabolism and fat deposition in the host.

Consistent with this interpretation, other data indicate that many members of the microbiome may influence a person's susceptibility to metabolic syndrome. An important line of research indicates that the microbiome can promote metabolic syndrome via its effect on the immune system. People with metabolic syndrome tend to display chronic low-level inflammation. The cause is a highly immunogenic component of the cell wall of many bacteria, known as lipopolysaccharide (LPS). LPS released from bacteria in the gut becomes associated with the low-density lipoproteins (LDLPs) which transport lipid from the gut to the blood and other tissues. The more fat in the diet, the more LDLP in the gut, and the more LPS is transferred to the bloodstream and internal organs. Individuals who are obese or have type 2 diabetes tend to have very high LPS levels in the blood. The resultant inflammation of internal organs is especially pronounced for the adipose tissue, which stores fat and is also a crucial source of hormones and other signalling molecules that regulate the body's metabolism. Chronic inflammation of adipose tissues impairs breakdown of fat droplets and promotes insulin resistance and type 2 diabetes, as well as cardiovascular disease and certain cancers.

Other studies have linked the activities of gut microorganisms to deleterious cardiovascular changes associated with metabolic syndrome. Gut bacteria convert carnitine, a constituent of meat, to the compound trimethylamine (TMA), which is transported via the bloodstream from the gut to the liver. Once in the liver, TMA is metabolized to trimethylamine-N-oxide (TMAO). TMAO modifies cholesterol metabolism in various ways, causing increased deposition of fatty material on the inner walls of arteries, thereby promoting atherosclerosis, a major cause of cardiovascular disease. Many microorganisms in the human gut microbiome are capable of producing TMA.

In the preceding paragraphs, I described some of the evidence suggesting that the Christensenellaceae and bacterial-derived LPS and TMA can influence human metabolic health. The experiments leading to these important discoveries are part of a much wider and very active research effort, with major new findings reported every month. It is becoming clear that the mechanisms are inescapably complex. Some groups of microorganisms and many microbial products can protect against metabolic syndrome, while other microbes and their products tend to promote disease, and the magnitude and direction of these effects can depend on the genetic make-up and age of the human host, as well as diet and exercise. The evidence that many microbiome factors have interconnected effects on human metabolism and its regulation is entirely to be expected because host functions have been influenced by resident microorganisms for vast periods of evolutionary time, back to our pre-animal ancestors (see Chapter 1).

Bariatric surgery and the restructuring of a dysbiotic microbiome

Bariatric surgery is a weight-loss surgical procedure. It involves the removal of a substantial part of the stomach, such that the patient can only take small meals. Some bariatric procedures also remove part of the small intestine, thereby reducing the uptake of nutrients. Generally, patients lose substantial amounts of weight over a period of one to several years, with associated reduction in other indices of metabolic syndrome. However, insulin sensitivity can often improve within days of the surgery, long before weight reduction is evident. In other words, bariatric surgery appears to have an anti-diabetic effect that is independent of the wider metabolic status of the patient.

There is now the strongest evidence that the rapidly improved insulin sensitivity following bariatric surgery is mediated via the gut microbiome. In a nutshell, the surgery causes a change in the microbiome to a composition that metabolizes bile acids differently, specifically favouring one compound (cholic acid 7-sulphate, abbreviated to CA7S), which has strong anti-diabetic effects.

Let us consider the mechanism in greater detail. The primary bile acids are produced by the liver and transferred to the small intestine (Figure 6(a)). These compounds are derived from cholesterol and, for many years, they were believed to be waste products of cellular metabolism with the incidental function of promoting lipid absorption from the small intestine. How wrong we were! It is now realized that bile acids are also potent signalling molecules; they regulate energy metabolism, including the levels of lipids and glucose which, when perturbed, result in metabolic syndrome. The efficacy of bile acid signalling is strongly influenced by the microbiome because many of the gut microorganisms transform the primary bile acids into a variety of chemically related compounds known as secondary bile acids. The pattern of microbial transformations varies between different people, according to the composition of the microbiome, directly affecting whole-body regulation of energy metabolism.

The shift in the composition of the gut microbiome following bariatric surgery is associated with a change in bile acid metabolism and profile of secondary bile acids (Figure 6(b)). In particular, the levels of one secondary bile acid, lithocholic acid (LCA), is elevated, and much of the LCA is returned to the liver

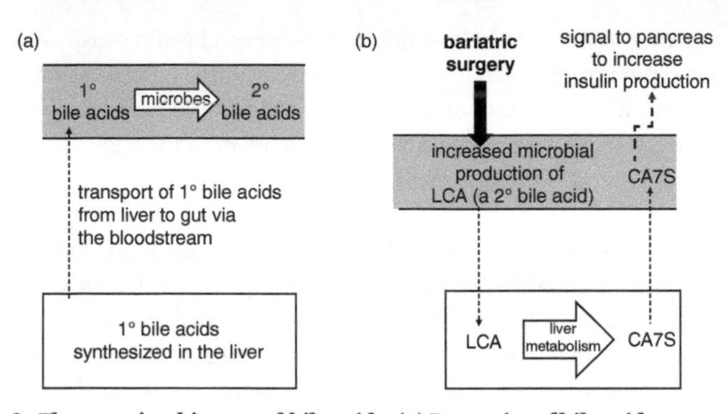

6. The gut microbiome and bile acids. (a) Dynamics of bile acid transformations in the human body. Primary (1°) bile acids (dominated by cholic acid and chenodeoxycholic acid and their conjugates with taurine and glycine) are synthesized in the liver and translocated to the gut. They are subsequently metabolized to various secondary (2°) bile acids by microorganisms in the gut. (b) Following bariatric surgery, changes in the microbiome, including reduced abundance of *Clostridium* species, result in a shift in microbial metabolism of bile acids to favour production of lithocholic acid (LCA). Much of the LCA is transported to the liver, where it is converted to cholic acid-7-sulphate (CA7S). CA7S is returned to the gut, together with other primary bile acids, where it functions as a signal, resulting in increased insulin production in the pancreas and improved insulin sensitivity.

via the bloodstream. In the liver, LCA is metabolized to CA7S, which is then returned to the gut, along with other primary bile acids. CA7S is a powerful signalling molecule, resulting in increased insulin production and, consequently, improved control over glucose and improvement of type 2 diabetes symptoms.

Importantly, the beneficial changes in bile acid metabolism and energy metabolism arising from bariatric surgery can be replicated by altering the composition of microbes without such surgery. The evidence for this comes from experiments on laboratory mice that were administered the bariatric procedure, with control mice receiving sham surgery. Six weeks later, the microbiome from the treatment and control mice was transferred to germ-free mice fed on a high-fat diet. After just two weeks, levels of LCA and CA7S were higher in the recipient mice (which had undergone no surgery) bearing the microbiome from the treatment group, relative to those with the microbiome from the control group. Other molecular indices in the treatment group indicated that these mice had improved insulin sensitivity.

The important follow-on task is to build on these mouse experiments to devise routes to modify the microbiome of human patients with metabolic syndrome to favour anti-diabetic function without any surgical intervention. I return to this issue in Chapter 7, where I address strategies that are being applied to shift a dysbiosis to a health-promoting microbiome.

Chapter 4
Microbiomes, the brain, and behaviour

'My microbes made me do it!' Until recently, this claim would have sounded like an excuse of last resort, and any suggestion that the gut microbiome may influence our behaviour would have been discounted as fanciful speculation. However, there are increasing indications that complex behavioural traits, including learning, memory, and emotional state, may be influenced by the composition and activities of the gut microbiome. Much of the evidence to date has been obtained from animal experiments, especially using the laboratory mouse and *Drosophila* fruit flies. Nevertheless, we should expect these results to be relevant to human behaviour, given the very long evolutionary history of interactions between animals and their resident microorganisms.

Behaviour, the brain, and the microbiome

Behaviour can be defined as the coordinated response of an animal to external or internal stimuli. For example, a cat may chase a mouse that it sees (response to an external stimulus), forage for mice when it is hungry (response to an internal stimulus), or only chase the mouse when it is hungry (combined response to external and internal stimuli). Many biologists and most psychologists would add a proviso to this definition: that behaviour has evolved to have consequences, to effect either a change in the relationship of the organism with its environment (the cat catches the mouse) or to maintain the status quo (the mouse escapes). Furthermore, behaviour does not necessarily involve movement. Planning your activities for the day may involve no movement but it has more significant consequences than, for example, blinking.

The biological substrate for animal behaviour is the nervous system, including the brain. The key functional unit of the nervous system is the neuron (also known as the nerve cell; a nerve is made up of many neurons). The nervous system transmits information very rapidly, by electrical signals along the length of each neuron and by chemical signals (neurotransmitters) released at specialized sites, called synapses, between neurons. The basis of any behaviour is a three-step process: sensory input, central integration, and the behavioural response, which usually involves motor output (Figure 7). The integration step can be highly intricate where the behavioural repertoire is flexible or complex, and it can involve large numbers of neurons in multi-way communication via synaptic connections that stimulate or depress the electrical activity of other neurons. In most animals, the principal site of integration is the brain. The human brain contains approximately 100 billion (10^{11}) neurons, each of which is estimated to have, on average, 1,000 synapses with other neurons.

There is now accumulating evidence that the gut microbiome can influence nervous system function, and particularly the central integration that dictates the behavioural response to internal and external stimuli. At first sight, this appears biologically far-fetched because the gut microbiome is remote from the brain. However, interactions between the gut microbiome and brain function are a simple extension to the long-recognized physiological relationship between the brain and gut, mediated primarily by the nervous system and hormones: the so-called gut–brain axis. We don't have to be physiologists to appreciate how emotional state, including stress, influences appetite and gut function, and how eating a nourishing meal can promote our sense of wellbeing.

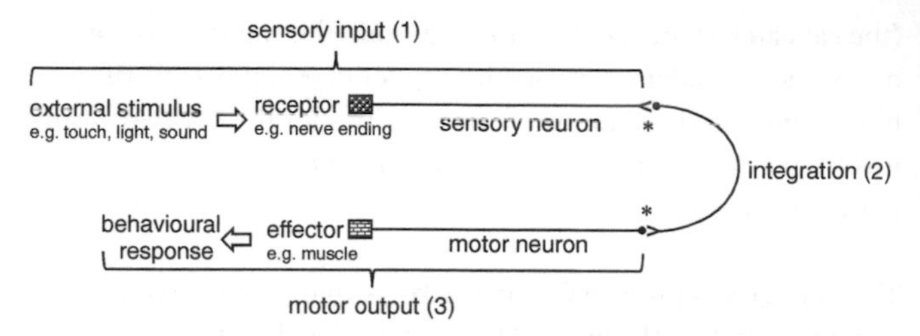

*Synapse: site for transfer of chemical signals (neurotransmitters) between neurons

7. The nervous system as the substrate for behaviour. The three-step process comprises: (1) sensory input, (2) integration of the sensory inputs and the internal state of the animal, leading to (3) a behavioural response, which usually involves movement of the animal, i.e. motor output (shown). Note that integration ranges in complexity from a single relay neuron (shown) to complex patterns of signalling that involve multiple regions of the brain and sophisticated evaluation of information, generating a coordinated behavioural response; and some behavioural outputs do not involve movement (see text for details).

The behaviour of germ-free animals

A powerful route to investigate the effects of resident microorganisms on behaviour is to study how removal of all microorganisms from an animal affects behaviour. Most of the research has been conducted on germ-free laboratory mice maintained in cages that are completely devoid of any microorganisms.

Mice are highly social animals, and they are inquisitive about their surroundings. Many of the studies have focused on these behaviours. In one experimental design, a test mouse was put into a chamber positioned between two other connected chambers, one empty and the other containing a second mouse. The test mouse is very likely to run to 'check in with' the second mouse, and to ignore the empty chamber. However, a germ-free mouse is just as likely to run to the empty chamber as the one with the mouse, indicative of low sociability. Another type of experiment made use of consistent differences in behaviour between two strains of laboratory mice. One strain was very timid; these mice hesitated for several minutes before jumping down from a narrow platform to explore an arena. A second strain, designated risk-takers, jumped down within 10–20 seconds. The behaviour of the mice shifted when they were colonized with the microbiome of the other strain: the risk-takers became more timid, and the timid mice became more adventurous.

Other studies have shown that germ-free mice can have impaired learning and memory, and be hyperactive or groom themselves for excessively long periods of time. But an important difficulty has emerged. The magnitude and sometimes even the direction of the differences between germ-free and control mice varies widely between studies. Some—but not all—of the differences can be attributed to the precise design of the experiments, whether male or female mice were used, and the choice of mouse strain. For researchers, this is very frustrating, but it does tell us that the behavioural effects of the microbiome are highly complex and can depend on many factors, some of which the researchers are not aware of.

Studies of other animals are also finding striking, but not always consistent, effects of eliminating the gut microbiome on behaviour. This is particularly evident for research on the fruit fly *Drosophila melanogaster*. As in mice, germ-free *Drosophila* can display impaired learning and memory, although the effects are subtle. Eliminating the microbiome also affects the activity of the flies; germ-free flies tend to be more active by day, and to sleep more at night. Effects relating to courtship behaviour and mating have also been reported. Germ-free flies are relatively unfussy about their mating partners, while courtship involving flies harbouring specific bacteria leads to mating much more frequently with partners bearing the same bacteria than with flies that bear no bacteria or different bacteria. However, as with so many behavioural effects of the microbiome, these fascinating observations are not reliably repeatable, and there are indications that the importance of the gut microbiome in mate choice by *Drosophila* may vary with the genetic make-up of the flies.

In summary, it appears that the microbiome can have substantial effects on various aspects of animal behaviour, including complex traits such as cognition and emotional state, but that these effects are not reliably displayed. The interactions are profoundly complicated and context-dependent. Realizing this, many researchers have focused on the underlying mechanisms. Their rationale is that understanding precisely how the gut microbiome affects a behavioural trait should provide clues to explain variable results, and also provide a rational basis for the biomedical application of these studies to address human neurological diseases, as discussed later in this chapter.

Microbial products and brain function

Germ-free mice, which have been so important in demonstrating the behavioural effects of the microbiome (see above), have also provided key information about the underlying mechanisms. The brain of a germ-free mouse is unhealthy in various ways. It is deficient in key proteins that support the survival and growth of neurons, and it has altered levels of many neurotransmitters. Anatomical differences are also evident, including the patterns of synaptic connections between neurons and overall morphology of several regions of the brain; these include the amygdala (important for processing threatening stimuli), hippocampus (with a major role in learning and memory), and thalamus (which relays sensory information to higher brain centres). It is unrealistic to predict precisely how these biochemical and morphological differences influence behaviour, but we can expect them to include emotional awareness, learning and memory, and the integration of sensory information.

How do microorganisms in the gut affect brain function? The simple answer is that microorganisms release chemicals that affect the activities of animal cells. Of particular importance are two types of released materials: components of the surface coverings, especially the cell wall, of microbial cells that are shed from both proliferating and dying cells; and waste products of microbial metabolism. (These products have already been considered in Chapter 3 because they also influence host metabolism.) Many of the released materials diffuse away from the microbial cells in the gut cavity (lumen). They are readily detectable in the gut wall, and they also pass into the bloodstream, which delivers them to distant organs of the body.

There is a more complicated answer to the question about how gut microorganisms influence brain function. It is that most microbial chemicals generally do not interact directly with the neurons in the brain. Instead, they affect the function of other animal cells, particularly cells in the gut wall that produce hormones (endocrine cells) and cells of the immune system, with downstream consequences for the brain and behaviour. In the next section, I introduce some of the patterns of interactions by addressing some widely held claims about how gut microbe–brain interactions affect the so-called happiness hormone.

The 'happiness hormone' serotonin

Serotonin is a small molecule that packs a big punch. It is best known for its role in supporting our feelings of wellbeing. Depression is associated with low serotonin levels in the brain, and various anti-depressant medications function to boost serotonin levels. The popular press and internet are awash with stories that the gut microbiome promotes happiness via its effect on brain serotonin levels. These stories are based on two fundamental misunderstandings.

The first misunderstanding is the erroneous claim that gut microorganisms produce most of the serotonin in our bodies. It is true that approximately 90 per cent of the serotonin in the body is synthesized in the gut. However, the dominant source of gut-derived serotonin is specialized endocrine cells in the gut wall. Although some gut microbes do release serotonin as a metabolic waste product, the amounts are very small.

The second misunderstanding is the incorrect belief that gut-derived serotonin (whether from microbes or endocrine cells) contributes directly to the levels of serotonin in the brain. It is true that serotonin made in the gut is delivered to the bloodstream, taking it to distant organs of the body. However, gut-derived serotonin is not delivered to the brain. The transfer of most blood-borne compounds, including serotonin, into the brain is severely restricted by a layer of cells known as the blood–brain barrier. The serotonin in the brain is synthesized in the brain.

Turning to scientific facts, we know with confidence that endocrine cells in the gut are stimulated to synthesize and release serotonin by various microbial products (e.g. indole, tyramine, and secondary bile acids have been implicated). We also know that endocrine cell-derived serotonin has important functions, but that these functions are not directly related to behaviour. It activates neurons in the gut wall, stimulating gut muscle to move food along the gut, and it is also released into the blood, where it promotes efficient blood clotting after injury, and supports healthy bone development and immunity.

So, to conclude, although the gut microbiome does not contribute directly to the serotonin levels in the brain, various behavioural and neurophysiological studies indicate strongly that the gut microbiome can contribute to emotional health. Some of the underlying mechanisms for this and other behavioural traits are addressed next.

The gut microbiome and the brain

There is growing evidence that the gut microbiome can influence behaviour in many different ways, by its effects on various physiological systems of the animal host, including the nervous system, immunity, and metabolism. The following examples serve to illustrate the diversity in these interactions.

The first and the most obvious route for gut microbiome communication with the brain is one that requires a neural connection between the gut and the brain. The primary candidate for a neural connection is the vagus nerve, which mediates two-way communication between the gut (and other organs, including the liver, lungs, and heart) and brain. The importance of the vagus nerve has been illustrated by experiments conducted on laboratory mice. In one study, mice harbouring lactobacilli bacteria displayed reduced anxiety and were more social than mice without lactobacilli. When the branch of the vagus innervating the gut was severed, these beneficial effects of lactobacilli in the gut were lost. The interaction between gut microorganisms and the vagus nerve is indirect; endocrine cells in the gut wall respond to fermentation products released from the lactobacilli by signalling to nearby nerve endings of the vagus.

A second important route by which microbes can influence the brain is via the immune system. Two linked processes are involved. First, the gut microbiome modulates immune responses, depending on its composition. Some gut bacteria activate pro-inflammatory responses, while others promote immune cells that are anti-inflammatory and foster immunotolerance. The second process is the extensive coordination between the immune system and nervous system. For example, some neurons can detect and respond to signalling molecules that regulate the activity of the immune system. The relationship between the immune and nervous system is also evident at the behavioural level: many animals, including humans, display non-specific sickness behaviour (lethargy, loss of appetite, high sensitivity to pain, etc.) in response to activation of the immune system, even in the absence of a pathogen, for example following some vaccinations.

Research on one type of immune cell, the macrophage, illustrates how the immune system can act as a go-between from the gut microbiome to brain function. Macrophages are mobile cells associated with many organs of the body, including the gut wall, and they provide protection against invasive microorganisms. The activity of a macrophage is regulated by signals from other cells. An important class of signalling molecules are the cytokines, small proteins produced by various cell types of the animal body that coordinate the activity of different parts of the immune system. Some members of the gut microbiome induce nearby cells to produce pro-inflammatory cytokines, which—in turn—stimulate macrophages to synthesize a chemical called kyneurenine (Figure 8). The macrophage-derived kyneurenine is released into the blood and it is one of a small number of compounds that readily cross the blood–brain barrier. Once in the brain, kyneurenine is transformed into neuroactive metabolites, including kyneurenic acid. The moderate levels of kyneurenine that reach the brain in most individuals are important for normal brain function. However, a gut microbiome that promotes inflammatory responses in the gut can lead to high concentrations of kyneurenine and its products in the brain, suppressing key neurotransmission events that underlie learning and spatial memory.

The gut microbiome can also affect behaviour through its influence on nutrition and metabolism. One route that is mechanistically trivial but behaviourally important concerns microorganisms which provide essential nutrients. Shortfall of a nutrient, whether arising from dietary or microbial deficiency, induces a specific hunger for that nutrient, leading to increased

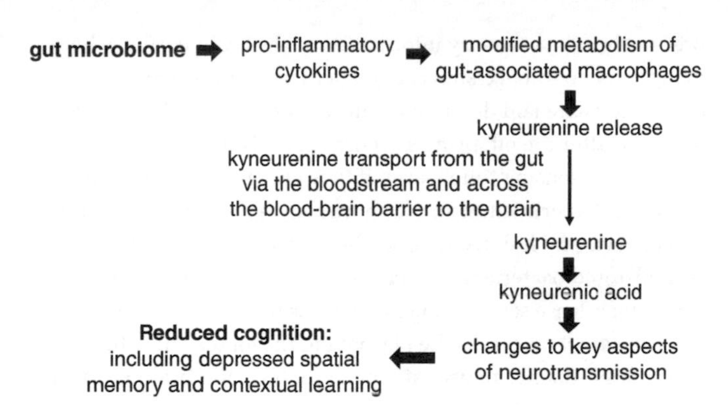

8. How modulation of gut immune cells by the gut microbiome can influence cognition in the brain. The composition of the gut microbiome varies among individual hosts, and a pro-inflammatory microbiome alters the metabolism of macrophages, which are immune cells that patrol the gut wall. The increased production of kyneurenine by macrophages under inflammatory conditions results in changes to brain chemistry, and impaired cognition.

foraging behaviour and often increased feeding or a shift in food choice. Other effects are mechanistically less obvious. For example, the presence and composition of the gut microbiome influence how much time *Drosophila* flies spend moving, and this is mediated by microbial effects on blood sugar levels, which in turn influence the activity of specific neurons in the fly brain that drive locomotion.

Finally, there is a very special route by which the gut microbiome can influence the brain of mammals: certain products of gut microorganisms can cross the placenta of pregnant females, affecting the development of the brain of the foetus and, ultimately, the behaviour of the offspring after birth. Much of the evidence derives from the detailed studies of the brain in germ-free mice, in which many neural circuits in the brain are altered. The anatomical perturbations are especially pronounced for the thalamus, a region deep in the brain which, as mentioned previously, relays sensory information to the brain cortex, where many behavioural decisions are made. These abnormal connections are laid down in the developing foetus of germ-free mothers; after the offspring are born, they display lifelong problems in sensing touch, even if they are provided with gut microbes. Experimental studies have shown that these anatomical and behavioural lesions are abolished when the mother is administered bacteria of one particular group, the clostridia. The clostridia release several small molecules that pass into the bloodstream, and across the placenta to the foetus. How these molecules influence neuronal growth in the foetal brain is under intensive study.

Much of the research on how the microbiome influences brain function and behaviour of laboratory animals, such as mice and flies, is driven by curiosity. But there is a second and pressing motivation: that this research may contribute to understanding and treatment of neurological diseases in humans. The 'smoking gun' for biomedical researchers is that the gut microbiome of patients with these diseases is commonly different from healthy people. Progress in research on the microbial dimension of these diseases involves extensive use of laboratory animals, including strains that are genetically modified to display disease symptoms similar to humans. For example, mice or flies can be modified to express a human gene that is strongly associated with a neurological disease.

In the remainder of this chapter, we will address how the combined study of humans and laboratory animals is providing strong indications that perturbation of the gut microbiome may contribute to neurodevelopmental disorders, particularly autism spectrum disorder (ASD), neurodegenerative diseases, including Alzheimer's and Parkinson's diseases, and mental health conditions, such as depression and anxiety disorders. These microbial perturbations, or dysbiosis, implicated in disease offer a potential route for microbial therapy.

The microbiome and neurodevelopmental disorders

Many human neurodevelopmental disorders are known, and most of them have a genetic basis, although the symptoms can be influenced by environmental factors. The severity of one disorder, however, may be linked to perturbation of the gut microbiome. This is autism spectrum disorder (ASD), with symptoms including poor social skills, repetitive behaviour, and, often, digestive complaints. The gut microbiome of children with ASD tends to differ from other children, although the condition cannot be linked reliably with an excess or deficit of any individual species or group of microorganisms. The correlation between ASD and altered gut microbiome composition could mean that the perturbed gut microbiome may contribute to the ASD symptoms. Alternatively, the perturbed gut microbiome could be a result of the gastrointestinal disorder, with no consequences for the development and function of the child's brain.

An indication that the gut microbiome may contribute to ASD comes from many reports that antibiotic treatment for unrelated ailments resulted in a temporary improvement in the behaviour of some (but not all) children with ASD. Another line of evidence has come from animal studies, notably a set of experiments designed specifically to test whether gut microorganisms can induce ASD-like symptoms in the laboratory mouse. Germ-free mice of a single strain were administered faecal samples from children with ASD or controls, referred to as neurotypical donors. The mice with the same type of microbiome were caged together, and the behaviour of their offspring was studied soon after weaning. To test for sociability, the test mouse was placed in an arena and, 10 minutes later, a second mouse was added. The behaviour of the test mouse over six minutes was videoed: did the test mouse interact with the second mouse, and for how long? The mice harbouring the ASD microbiome were significantly less sociable than the mice with the neurotypical microbiome. The second assay tested for obsessive behaviour. Making use of the tendency for mice to bury novel objects in their bedding, each mouse was placed in a box with 20 marbles evenly placed on top of a layer of wood chips, and the number of marbles buried over a 10-minute period was scored. Most of the mice with the microbiome from neurotypical donors buried five or fewer marbles, spending most of the time investigating their new surroundings. The mice with the ASD microbiome behaved differently; after encountering the first marble, many of them spent the remainder of the 10-minute trial engaged in the highly repetitive behaviour of burying marbles.

This study is important because it demonstrates that the microbiome of children with ASD can induce behaviours in mice that are a reasonable proxy for the key behavioural symptoms of ASD. Nevertheless, we should be cautious in translating these studies back to humans. Behavioural assays developed for the laboratory mouse, such as the direct social interaction assay and the marble burying assay described in the previous paragraph, are well suited to mouse biology and provide valuable contributory evidence, but they cannot possibly capture the complexity of human behaviour. The value of this research is that it provided a rational basis to study the microbiome–host interactions that underlie the ASD-like behaviour in mice bearing a microbiome derived from an ASD donor. These follow-on studies have led to the identification of two products of bacterial metabolism, taurine and 5-amino valeric acid, that alleviate ASD-like symptoms in mice. These metabolites offer leads for new treatments for children with ASD.

The microbiome and neurodegenerative disorders

Neurodegenerative diseases arise when large numbers of neurons fail to function and die. The affected neurons may be in the brain or in the peripheral nervous system, including neurons receiving information from sensory organs or delivering information to the muscles of the body (see Figure 7). The causes of neurodegenerative diseases are complex, often with contributions from genetic and environmental factors. The gut microbiome has also been implicated in two of the most common neurodegenerative diseases: Alzheimer's disease, in which progressive memory loss (also known as dementia) is associated with the abnormal aggregation of proteins, known as amyloid plaques and neurofibrillary tangles, in the brain; and Parkinson's disease, caused by degeneration of neurons in a region of the midbrain, the substantia nigra, which is important in the coordination of voluntary muscle activity and movement.

As for autism, the gut microbiomes of patients with Alzheimer's disease and Parkinson's disease differ from healthy individuals of similar age. Also as for autism, these difference may contribute to, or be a consequence of, the disease. Intriguingly, many patients with Parkinson's disease have experienced disorders of the digestive system, often for many years before the onset of neurological symptoms, and there are indications that the Parkinson pathology may start in the gut, with damage to neurons associated with the gut wall preceding degeneration of neurons in the brain.

Strains of laboratory mice that are predisposed to display the neurological symptoms of Alzheimer's disease or Parkinson's disease suggest that the altered gut microbiome may play a role in these diseases. When these mice are administered microbiome samples from mice or humans with the disease, they suffer worse symptoms than genetically equivalent mice provided with microorganisms from healthy donors. The microbiomes of mice with Alzheimer-like or Parkinson-like pathology tend to be enriched in bacteria that stimulate immune responses and promote inflammation. Similar functional differences in the microbiome are also evident between the gut microbiome of humans with these diseases versus healthy humans.

These findings are leading some researchers to suggest that Alzheimer's disease and Parkinson's disease may not be disorders exclusively of the brain, but also involve whole-body immunity, especially the immunological interactions with the gut microbiome. Intriguing evidence is coming from laboratory research. One promising lead comes from *Drosophila* flies that are genetically modified so that they are likely to develop Alzheimer-like symptoms, including amyloid plaques in the brain and impaired memory. This predisposition for disease is greatly exacerbated by a gut bacterium, a strain of *Erwinia carotovora*. The underlying mechanisms are partly understood. *E. carotovora* in the gut induces a change in the immune system, causing macrophage-like immune cells to migrate to the brain. In this location the immune cells trigger inflammation, resulting in the death of many neurons, especially in the region of the fly brain responsible for learning and memory. A further genetic modification that eliminated the macrophage-like immune cells abolished the bacterial induction of Alzheimer-like symptoms. Of course, these experiments cannot be applied directly to develop treatments of human patients for many reasons; *Erwinia* bacteria are rare or absent from the human gut microbiome, and there are major biological differences in immunological and brain function of *Drosophila* and humans. The importance of these experiments is that they alert researchers conducting experiments on mice and treating human patients to the possibility that the gut microbiome can change the properties of immune cells, resulting in inflammation of the brain and the development of neurodegenerative diseases.

Mental health, including depression and anxiety

Clinical interest in the role of the microbiome in mental health disorders, especially depression and anxiety, has arisen from the finding that the microbiome of affected individuals is often different from healthy people. Animal models provide supportive evidence. For example, germ-free mice display depressive-like behaviour when administered microbiome samples from people with depression; and germ-free mice display anxiety-like behaviour. Furthermore, administration of specific bacteria, including lactobacilli and *Bifidobacterium*, can ameliorate both depressive and anxiety-like behaviour in mice. The positive effects of these bacteria have been attributed to gut-to-brain signalling via the vagus nerve (as discussed earlier in this chapter). Evidence for other mechanisms independent of the vagus has been obtained; for example, these disorders have also been linked to microbial production of a specific fermentation product, butyric acid. These findings indicate that there is probably much variation in whether and how gut microorganisms influence depression and anxiety-related symptoms, reinforcing the wide recognition that the mechanisms underlying these disorders can be diverse.

Translating these microbiome discoveries made from animal studies back to the mental health of humans is even more fraught with difficulty than for neurodevelopmental and neurodegenerative diseases. This is because self-reporting by the patient plays a key role in diagnosis and monitoring of recovery from mental illness. Self-evidently, animal studies are dependent on other criteria. For example, a very widely used assay for depression-like behaviour in mice is the 'tail suspension test'. A mouse is suspended by the tail just above the ground for five minutes (this is not painful for a mouse) and the amount of time during which it tries to escape is scored. A mouse that becomes immobile soon after the start of the test is interpreted as despairing of escape and so 'depressive'. Although this type of assay cannot encompass the complexity of the human disorder, it can provide valuable information. For example, biochemicals that increase the time that the mouse struggles in the tail suspension test have gone on to become valuable medications for depressive illness in humans. Researchers are still discovering whether these assays in mice yield results for effects of microbiome composition that can, similarly, be applied to the human disorder.

The knowns and unknowns of microbiome effects on behaviour

Although the data are still fragmentary and sometimes contradictory, there is a strong signal that the microbiome can influence brain function and behaviour, and that a perturbed microbiome may contribute to several common and debilitating neurological diseases in humans. However, many unknowns remain. It is still uncertain how readily a dysbiosis can be corrected in humans, and the physiological and psychological consequences of such interventions remain unclear. For now, there is no scientific basis to claims that simple microbial therapies, such as swallowing a capsule of bacteria or daily consumption of probiotic yogurt, might resolve the symptoms of these diseases.

Only time—and much hard work by microbiome researchers—will tell whether microbial therapies will play an important role in the treatment of neurological disease. But an important achievement already obtained by this research area is that it has changed the conversation. We can understand the healthy development and function of the brain, and brain disorders, only by considering the biology of the entire body, including our microbial partners.

Chapter 5
Microbiomes and infectious disease

We live in a dangerous world. From the perspective of a microorganism, an animal or plant is a large patch of concentrated nutrients. Some bacteria, microbial eukaryotes, and viruses exploit this nutritious patch aggressively, leading to disease of the animal or plant. The sustained health of animals and plants depends partly on effective defences that resist infection by these pathogens, and partly on resident microorganisms which provide complementary protection.

In the greater part of this chapter, we will explore how the microbiome influences host susceptibility to infectious disease. A major theme is the role and mechanisms underlying the protective function of the microbiome, often in close collaboration with the immune system. However, the alliance between resident microbes and host sometimes breaks down, such that members of the microbiome can become pathogenic or exacerbate infectious disease. In the closing sections, we turn to a different topic of critical importance for public health: disease agents that are transmitted by insect vectors.

Colonization resistance

The widespread medical use of antibiotics since the 1940s has transformed the treatment of infections caused by microbial pathogens. However, an important clinical problem became apparent very quickly: an unexpectedly large number of patients become infected by an unrelated pathogen in the days following cessation of the antibiotic treatment. Researchers discovered that this problem could be reproduced very readily in laboratory mice for a diversity of different pathogens and types of antibiotic. For example, one early study discovered that mice were up to a million times more susceptible to the pathogenic bacterium *Salmonella enterica* if they had previously been administered the antibiotic streptomycin. Furthermore, the mortality of mice challenged with a pathogen was particularly high following antibiotic treatments that severely depleted or eliminated the resident microorganisms in the gut, and the resistance of the antibiotic-treated mice to the pathogen was restored by providing them with gut microbes from mice that had not been administered antibiotic. These experiments showed that members of the gut microbiome drastically reduce the susceptibility of mice—and, presumably, humans—to pathogens. This protective effect of the microbiome against pathogens has come to be known as colonization resistance.

Although the role of the resident microorganisms in protecting the host from pathogens was first demonstrated for gut microorganisms, this function of the microbiome is general. For example, resident microorganisms in the skin help to defend intact skin from attack by pathogens, such as *Staphylococcus aureus* and *Streptococcus pyogenes*, and can also prevent infection of wounds and promote healing. Similarly, some members of the oral microbiome suppress the bacteria that cause tooth decay, and the lactobacilli that dominate the vaginal microbiome of many women protect against *Candida* fungal infections that cause vaginal thrush.

Experimental studies on a wide variety of animals, from chickens to insects, nematode worms, and even the simplest of animals, such as freshwater hydra polyps, recapitulate the studies on laboratory mice: that depleting the microbiome, whether by antibiotic treatment or other methods, leads to increased susceptibility to pathogens. This can be of practical importance. For example, beekeepers routinely administer antibiotics to their hives, either prophylactically or in response to the first symptoms of disease. Unfortunately, these antibiotic treatments can decimate the gut microbiome that protects bees against secondary infections. The negative consequences of antibiotics are often not evident to the beekeeper because infected bees tend to leave the nest to die. It has been suggested that the widespread use of antibiotics in recent decades has contributed to the dramatically increased incidence of colony collapse disorder, in which most of the bees disappear, over a few days, from a hive that is well provisioned with food.

Colonization resistance is also widespread in plants. By far the best exemplar is disease suppressive soils. The soil-derived microbiome associated with the roots of plants grown in these soils, the rhizosphere microbiome (see Chapter 2), defends plants against a range of pathogens, both soil-borne pathogens and also pathogens that infect the leaves and stems of the plant. Disease suppressive soils vary widely in extent (some are limited to a single field or orchard), and the protective function can be highly specific to one pathogen or it can be relatively general. Different soils with highly similar disease suppressive function can contain very different microbial communities, suggesting that microbial protection might be mediated by a wide diversity of microorganisms. Nevertheless, specific bacteria have been isolated from suppressive soils (e.g. *Streptomyces* and *Pseudomonas* species) and demonstrated to protect crops against agriculturally important pathogens, including *Rhizoctonia* fungi that cause root rot and blight of various crops and the fungal agent of take-all disease in cereals.

Colonization resistance as microbial self-interest

The microorganisms that display colonization resistance are not acting out of selfless concern for the wellbeing of the host. They are defending their host 'habitat' against intruders that would otherwise usurp their space and compete for host-derived nutrients, and these intruders include pathogens of the host. To a large extent, colonization resistance is just a by-product of the self-interested activities of resident microbes.

How do host-associated microorganisms defend themselves against competing microorganisms? Many members of the microbiome have an arsenal of chemical weapons, by which they kill or suppress the growth of competitors (Figure 9(a)). Some bacteria release toxins, known as bacteriocins, that kill only closely related bacteria, which are likely to have similar habitat requirements and so be the strongest competitors. This trait helps resident strains to persist, thereby contributing to the stability of

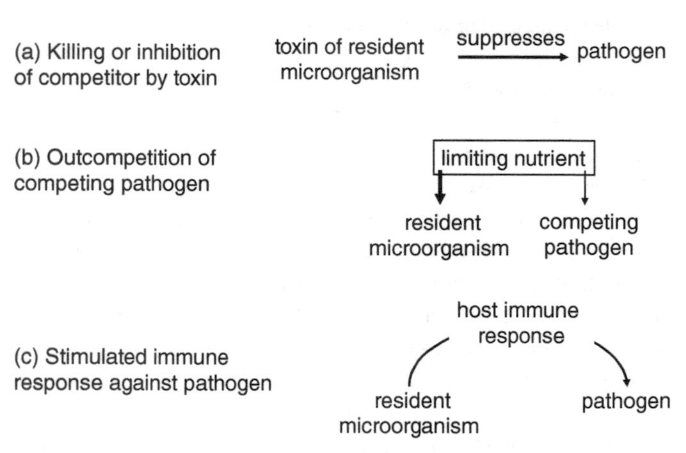

9. Mechanisms of colonization resistance: how resident microorganisms can protect their host against pathogens. (a) Direct interaction between a resident microorganism and pathogen mediated, for example, by a toxin that suppresses the colonization or growth of the pathogen. (b) Utilization of a limiting nutrient by a resident microorganism. As a result, the pathogen is nutrient-limited and fails to thrive in the host. (c) Stimulation by a resident microorganism of a host immune response that is more harmful to the pathogen than to itself.

the microbiome. These bacteriocin-producing bacteria are potentially important in disease prevention because they may limit the spread of antibiotic-resistant strains.

Some toxins produced by microorganisms are effective against a diversity of other microorganisms. For example, many by-products of fermentative metabolism of various bacteria in the human colon inhibit the proliferation of various pathogens, such as *Salmonella typhimurium* and *Clostridium difficile.* Antimicrobial compounds implicated in some disease suppressive soils include familiar antibiotics, since many of the medically important antibiotics were originally derived from soil microorganisms, such as *Penicillium* and *Streptomyces.* Other antimicrobial compounds, especially those produced by members of animal microbiomes, are biochemically novel and the subject of active investigation. These chemicals have potential as alternatives to current antibiotics.

Microorganisms associated with humans and other animals also protect their habitat by being effective competitors for space and nutrients (Figure 9(b)). Competition for nutrients can be intense in the human colon. Some bacteria degrade the mucus that is sloughed off from the gut wall, releasing specific sugars that are rapidly consumed by other bacteria. If the latter bacteria are depleted by antibiotic treatment, the concentration of these sugars increases, providing a resource for the rapid proliferation of pathogens, such as *S. typhimurium* and *C. difficile.* Similarly, many disease suppressive soils bear a dense and diverse microbial community that is believed to lock away soil nutrients, so starving out soil-borne pathogens.

Competition for resources can be highly specific. This has been demonstrated, for example, in studies of an insect-borne bacterium, *Wolbachia. Wolbachia* can suppress certain viral pathogens by limiting the uptake of virus particles into insect cells, required for viruses to proliferate. The antiviral properties of

Wolbachia can be attributed, at least partly, to its lifestyle. Unlike the bacteria associated with the gut, *Wolbachia* lives inside insect cells, where each *Wolbachia* cell is bounded individually by a lipid-rich membrane synthesized by the insect host. This lifestyle consumes large amounts of insect lipids, leaving a shortfall of lipid and membrane required for virus uptake.

In summary, some of the key routes by which resident microbes defend their host against pathogens involve toxin production and efficient resource acquisition. These traits are not peculiar to host-associated microbes; microorganisms in all habitats compete for resources. The special aspect of microbial interactions within a host is that the host can shift the balance of among-microbe interactions to favour beneficial microorganisms. For example, the host can provide specific nutrients that only beneficial microbes can utilize or display a muted immune response to beneficial microbes. The flip-side of this proverbial coin is that the resident microbes can also modulate the immune system to their advantage. One way that this is done is by amplifying the host immunological response to pathogens.

Microbial modulation of the host immune system

The microbiome is an essential partner in the regulated development and function of the animal immune system. Germ-free mice are very susceptible to bacterial infections. This is partly because they lack protective microorganisms that compete with the pathogen (see above). But there is another contributory reason. Germ-free mice have fewer 'first responder' immune cells, such as macrophages, which attack invading bacteria within the first few hours of infection. The problem is developmental: that the final stages in the development of these critically important immune cells generated in the bone marrow depend on chemical signals that are released from the gut wall in response to gut microbes. When germ-free mice are provided with gut microorganisms, the final stages in immune development can proceed. However, some aspects of immune function can develop properly only if the mice are provided with gut bacteria soon after birth, suggesting that the immune system is fully responsive to the microbial signals only during a narrow developmental window.

The microbiome also influences the overall level of defence-readiness of the mammalian immune system. This is partly dictated by the composition of a class of immune cells known as T cells. Some T cells are pro-inflammatory, promoting strong protection against pathogens, while other T cells are anti-inflammatory, favouring immunotolerance. The microbiome influences the relative abundance and activity of these different types of T cells (Figure 10). Regulatory T cells (Tregs) produce signalling molecules that dampen the immune response, and they are promoted by butyric acid and related products released from gut microbes as they ferment sugars and related compounds. The microbes with fermentative metabolism are intolerant of oxygen and live at a distance from the oxygen-rich gut wall. Other

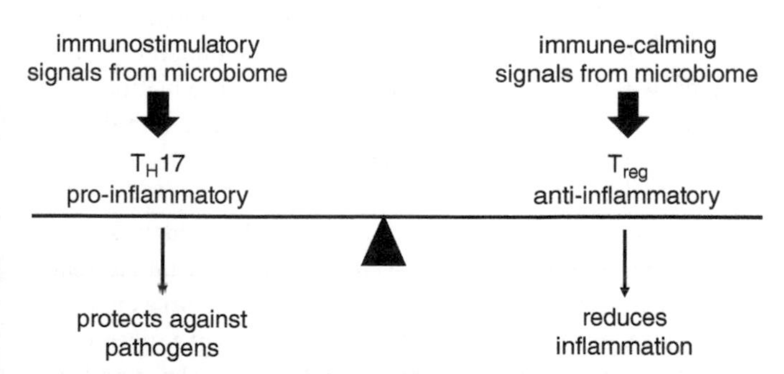

10. The microbiome and immunological balance. Immunostimulatory and immune-calming signals favour different arms of the immune system, as illustrated here for two types of immune T cells (T_H17 and Treg cells). An immune environment that is too stimulatory causes inflammation which can damage host tissues and organs, but a host with a subdued immune response is highly susceptible to pathogens.

bacteria, however, prefer the oxygen-rich conditions close to the gut wall. They do not release 'calming' molecules, such as butyric acid, and therefore do not stimulate Tregs. Instead, constituents of their cell walls and other products stimulate pro-inflammatory T cells. Similar interactions between the microbiome and immune system occur in the skin. The colonization of hair follicles by skin bacteria, such as *Staphylococcus epidermidis*, leads to the accumulation of Treg cells that promote immunological tolerance of non-pathogenic bacteria. Perturbation to this interaction can cause inflammatory conditions, resulting in acne and other skin disorders.

The continual communication between resident microorganisms and the host immune system provides the basis for an important type of colonization resistance: microorganisms protect their host by stimulating an immune response that is more damaging to the pathogen than to themselves (Figure 9(c)). A well-studied instance relates to an antimicrobial protein known as RegIIIγ, which is produced in the small intestine of mice. This protein literally punches holes into the outer covering of certain types of bacteria, including *Listeria*, an agent of food poisoning, and some strains of *Enterococcus*, which causes life-threatening infections in people with a weakened immune system. The production of RegIIIγ is dependent on the activation of immune cells in the gut wall by the resident gut microorganisms.

To understand why RegIIIγ is not harmful to the resident gut microbes, we need to consider the layer of protective mucus that lines the gut wall. Mucus is a biochemically very complex material with large quantities of polysaccharides, made up of a diversity of sugars. Some gut microorganisms live in the mucus layer and, as mentioned above, they break down the mucus, releasing sugars and other nutrients. However, mucus is also structurally complex. In most regions of the gut, it forms two distinct layers: an inner layer (next to the gut wall) which is very thick and is not easily degraded by bacteria; and an outer layer which is more fluid and much more readily utilized by bacteria. The resident microorganisms live, feed, and often swim around in the outer mucus, and some of the products they release, including discarded cell wall material, is detected by cells of the gut wall, inducing them to produce RegIIIγ. The RegIIIγ molecules get trapped in the thick, inner mucous layer. This is of no consequence for the resident, beneficial microorganisms in the outer mucus, but it is a major difficulty for pathogens. Pathogens use the oxygen that is available close to the gut wall to derive sufficient energy to proliferate rapidly and some pathogens then invade the gut wall to reach the bloodstream and other organs for a systemic infection. Essentially, the resident microbes induce the host to create a RegIIIγ-rich no man's land of inner mucus, thereby protecting both themselves and the host from invasion by pathogens.

I have considered in some detail how gut microorganisms stimulate production of the antimicrobial protein RegIIIγ in the mouse because this interaction has been studied in great detail. There is also good evidence that the equivalent protein in humans functions in the same way. This is just one of many routes by which resident bacteria interact with the host immune system, affecting host susceptibility to infectious diseases. Many of the underlying mechanisms are irreducibly complex, and interconnected, comprising a web of interactions that involve many signalling molecules of host and microbial origin, various types of immune cells, and a broad spectrum of antimicrobial compounds. Understanding of these microbe–host interactions holds great promise to deliver improved treatment of infectious disease and immune-related disorders, often involving multiple members of the microbiome. However, to achieve this goal, we need also to take into account a further issue: that not all members of the microbiome are beneficial to the host.

Hey microbe, whose side are you on?

In the opening sections of this chapter, I have argued that it is in the self-interest of resident microbes to 'look after' their host habitat, including to protect the host from pathogens. Nevertheless, this beneficial outcome is far from assured. The bacterium *Neisseria meningitidis* lives in the nose and throat of 10–30 per cent of the human population, where it does no harm and probably contributes to the important function of colonization resistance against pathogens. But very occasionally, *N. meningitidis* escapes from this location into the internal tissues of the body, where it can cause meningitis, a life-threatening inflammation of the membranes that bound the brain and spinal cord. The disease-causing bacteria are in the wrong place, both for the human patient and for themselves: it is much better for *N. meningitidis* to be in the nose of a healthy host, from where they can be transmitted via respiratory droplets to other people.

Neisseria meningitidis resembles Dr Jekyll, the pillar of society who was occasionally transformed into Mr Hyde, the personification of evil, in the famous story by Robert Louis Stevenson. Various other Jekyll-and-Hyde microbes inhabit our bodies. We have already noted the example of *Helicobacter pylori*, one of a few species of bacteria that can thrive in the highly acidic stomach, which is renowned for causing stomach ulcers in middle age, but which also promotes the development of a healthy immune system in children, reducing the incidence of asthma and other immune-related diseases. There is also growing evidence that some bacteria, such as *Staphylococcus epidermidis* and *Cutibacterium acnes*, which protect the healthy skin against pathogens, can colonize chronic wounds, such as skin ulcers, where they suppress normal healing.

The outcome of host–microbial interactions can also be influenced by phage. A telling example comes from research on *Pseudomonas aeruginosa*, a bacterium that can be a harmless member of the microbiome or a virulent pathogen, varying with strain and the health status of the host. Some aggressive strains of *P. aeruginosa* are infected with a phage that does not kill the bacterial cells but adopts a low profile, integrated into the bacterial DNA (the genetic material), so that it is retained as the bacterial cells proliferate. In some situations, these phage-colonized bacterial cells can attach strongly to mucus in the lung, and this promotes their persistence and spread through the lung, creating a chronic infection. Other studies on *P. aeruginosa* infections of wounds found that the phage-colonized cells suppress host immunity, and this prevents clearance of bacteria from the wounds and delayed wound healing.

'Every which way' for virus–microbiome interactions

Further insight into the variability in microbiome-mediated colonization resistance comes from research on pathogenic viruses. All organisms live under the threat of viral infection. A single virus particle can take over a living cell, forcing it to make vast numbers of progeny virus particles that can then infect neighbouring cells. Unsurprisingly, organisms possess antiviral defences which protect against many viruses and promote recovery from viral infection.

The immune system plays a central role in the protection of humans and other animals from viruses, and there is increasing evidence that the microbiome can modulate antiviral immunity.

Viral pathogens that infect host cells of the gut are known as enteric viruses. Various lines of evidence indicate that the gut microbiome can 'aid and abet the enemy', promoting enteric virus infections. Many studies have focused on noroviruses, which cause acute gastrointestinal infections (most notoriously on cruise ships), and the rotaviruses that are a major cause of diarrhoea, especially in young children. Definitive evidence that the gut microbiome facilitates viral infection comes from mouse experiments. Disease symptoms and viral load are much reduced in germ-free and antibiotic-treated mice infected with norovirus or rotavirus, and when these mice are colonized by microbes from untreated mice, their susceptibility to these viruses increases.

Gut microorganisms promote enteroviruses in several ways. The virus particles are prone to degrade in the complex chemical environment and warmth of the gut, but they are stabilized when they attach to the surface of some gut microbes; this does not appear to affect the bacteria. The gut microbes that feed on the mucus lining the gut wall can often reduce the thickness of the mucous layer, facilitating viral access to the gut cells. (Recall that the virus particles can only reproduce inside cells, and that the enteric viruses infect animal cells, but not bacterial cells.) In addition, the tendency of the gut microbiome to dampen immune responses (see above and Figure 10) includes suppression of key antiviral functions, particularly signalling via interferons, thereby facilitating the spread of viral infections.

These studies lead to the temptation to consider antibiotics as a treatment to suppress bacterial enhancement of enterovirus infection (antibiotics have no direct effect on viruses). There are two excellent reasons to resist this temptation. The first is that the deleterious effects of antibiotic-mediated perturbation of the microbiome can be greater and more persistent than the viral infection. The second is that other studies, particularly conducted on humans, provide tantalizing, albeit inconclusive, evidence that members of the gut microbiome can be protective against enteroviruses. For example, up to 30 per cent of the human population are infected with norovirus without any ill-health. These asymptomatic carriers tend to be distinctive, both genetically and in their gut microbiome composition, which includes a high abundance of *Bacteroidetes* bacteria.

The contradictory data concerning the effect of gut microorganisms on enterovirus infection tell us that the three-way interaction between the virus, gut microbes, and host immune system is very complex. Many factors can contribute to the different outcomes, including the composition of the microbiome, the status of the immune system, and other host-related variables that are shaped by genetic and environmental factors. The fragments of evidence that certain gut microbes may be protective against enteroviruses offer the opportunity to understand and manage the microbiome for improved health outcomes of patients that are especially vulnerable to these viral diseases.

The gut microbiome and infectious diseases of the lung

While various mouse studies are indicative of microbial promotion of enteroviral infections, equivalent experiments conducted with viruses that infect other parts of the body tend to yield the opposite results. In particular, the gut microbiome can be protective against infectious respiratory diseases. This interaction is often referred to as the 'gut–lung axis', analogous to the gut–brain axis considered in Chapter 4. It is believed that the gut microbiome modulates the immunological status of the lungs, primarily by influencing the abundance and activities of immune cells in the lungs.

The key evidence, yet again, comes from mouse studies. When mice are administered antibiotics that deplete the gut microbiome, they become very susceptible to influenza virus, linked to suppressed production of interferons and other antiviral compounds. It appears that specific compounds released from bacteria in the gut stimulate antiviral immunity throughout the body. One of these products is desaminotyrosine, generated by the metabolism of dietary flavonoids by some *Clostridium* bacteria; when desaminotyrosine was added to the drinking water of antibiotic-treated mice, the mice had higher levels of interferon and were better protected from the influenza virus.

These findings raise the important issue of whether the gut microbiome may influence the susceptibility and pathology of the SARS-CoV-2 virus which causes COVID-19. At the time of writing, definitive data are not available, but there are two lines of circumstantial evidence. The first is that the principal predictors of COVID-19 disease severity include obesity, type 2 diabetes, and hypertension, and all of these conditions are associated with dysbiosis of the gut microbiome (see Chapter 3). Second, the microbiome of patients with severe COVID-19 symptoms is dysbiotic, with many bacterial species that induce inflammatory response in the gut and a deficit of bacteria that calm immune responsiveness. Well-designed trials are providing evidence that probiotics may reduce the severity of disease symptoms of COVID-infected individuals. However, more extensive trials are essential, to identify how efficacy may be influenced by the microbiome composition and genetic make-up of the patient.

Potentially, a further set of microbial players may influence susceptibility to infectious respiratory disease: the lung microbiome. Many (but certainly not all) researchers judge that the lung microbiome is not important under most circumstances. The argument is that microorganisms detected in the healthy lung are mostly transient. Microbial cells from the resident microbial communities inhabiting the mouth and nose are frequently aspirated into the lungs as we breathe, but they fail to persist. However, the lung microbiome is not invariably a bystander in interactions between respiratory viruses and the host. Infections by some respiratory viruses, including the influenza virus, alter the immunological status and available nutrients in the lungs, facilitating persistent secondary infection by bacterial pathogens, such as *Streptococcus pneumoniae* and *Pseudomonas aeruginosa*. In this situation, the host alliance with the antiviral gut microbiome is in direct conflict with an alliance between the virus and lung-infecting bacterial pathogens, and the recovery of the patient can depend on the outcome of this conflict.

Summarizing so far, the microbiome is generally protective against infectious disease, and this protective function is mediated by a diversity of processes, including chemical warfare and starving out the invading pathogen, as well as modulation of the host immune system. However, the resident microbes are operating from self-interest, which is not invariably aligned with the interests of the host. Resident microbes can occasionally cause disease (Jekyll-and-Hyde), and some can promote pathogen infection (aid and abet the enemy). Although the interactions can be very complex with many variables, our developing understanding is already providing routes for microbial therapies to resolve otherwise intractable infectious diseases (see Chapter 7).

Vector-borne diseases

We have, so far in this chapter, focused on disease agents that susceptible hosts acquire from the environment, or by direct contact with infected organisms, and, for animals, via food and drink. Now let's turn to a further route of disease transmission: via vectors, i.e. living organisms that carry the pathogen from one host to another.

Some of the most important disease agents of humans, both historically and today, are vectored by insects and other arthropods. For example, mosquitoes vector the malaria parasites (*Plasmodium* species) and various viruses (e.g. dengue, zika); ticks vector the spirochaete bacteria that cause Lyme disease and relapsing fever; and fleas vector the bacterial agents of typhus and bubonic plague (*Rickettsia typhi* and *Yersinia pestis*, respectively). Many plant diseases are also vectored by insects. These include viral diseases of huge economic importance, such as rice stunt viruses, barley yellow dwarf viruses, cassava mosaic viruses, as well as bacterial pathogens, such as *Liberibacter* agent of citrus greening disease and *Xylella* agent of Pierce's disease of grapevines. Traditionally, the chief strategy to control diseases vectored by insects has been to target the vector, for example by insecticide sprays and habitat manipulation (e.g. draining standing water where many mosquitoes breed). There is, however, a great need for new approaches, especially because the incidence of insecticide resistance in many insect vectors is increasing.

One promising new technology for insect vector control is based on a single member of the microbiome, the bacterium *Wolbachia*, which occurs in up to 60 per cent of all insect species. As we noted earlier in this chapter, *Wolbachia* lives inside insect cells and suppresses viral infections of the insect host. This antiviral capability applies to viruses that are vectored by insects, as well as insect-specific viral pathogens. However, there is a second attribute of *Wolbachia* that is also crucial for its application to combat various insect-vectored diseases. This is its effect on insect sex and reproduction.

Wolbachia and insect sex

One way to describe *Wolbachia* is as a reproductive parasite. It can thrive in many organs of the insect body, but it has a special affinity for the ovaries of female insects. Here, *Wolbachia* inserts itself into the developing eggs, ensuring its transmission to all the offspring of its female host. This trait creates a conflict between the insect and *Wolbachia*. Under most circumstances, the ideal offspring sex ratio for the insect is an equal number of males and females, but *Wolbachia*, being transmitted exclusively via females, benefits from an excess of female offspring. *Wolbachia* resolves this conflict to its advantage by manipulating the sex ratio towards female offspring. In a few insects, *Wolbachia* kills or feminizes male offspring, but the most widespread mode of sex ratio distortion—and the mechanism that is being harnessed for insect vector control—is cytoplasmic incompatibility.

The term cytoplasmic incompatibility was coined to describe how crosses between certain strains of mosquitoes were infertile. Fertility could be rescued when a drop of the egg contents from females that were compatible with males of a certain strain was transferred into females of an incompatible strain (Figure 11). It was subsequently shown that cytoplasmic incompatibility was associated with *Wolbachia* living in the cells of the insect host, and it could be abolished by antibiotic treatment. We now know that a toxin produced by *Wolbachia* accumulates in the sperm of male insects and kills the egg just after fertilization, unless the toxin is countered by an antidote, synthesized by the *Wolbachia* in the egg.

Fortunately, *Wolbachia* does not infect vertebrate animals, and reproductive parasites with traits similar to *Wolbachia* are unknown in humans.

Returning to insect vectors, cytoplasmic incompatibility provides the route to drive *Wolbachia* to high levels in an insect population.

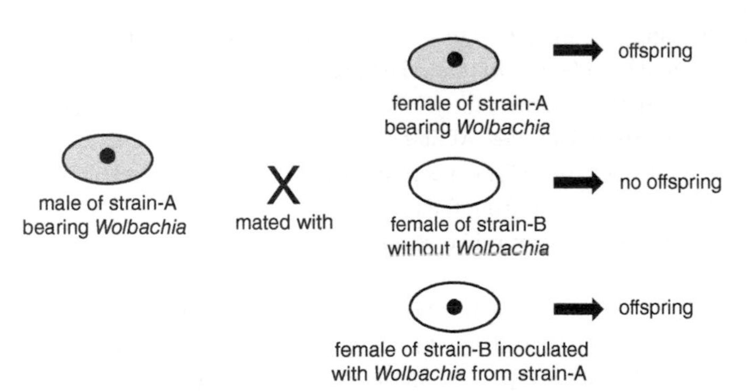

11. Cytoplasmic incompatibility. Crosses between male insects bearing *Wolbachia* and *Wolbachia*-free female insects are infertile. The reverse cross of a *Wolbachia*-bearing female and *Wolbachia*-free male (not shown) is fully fertile. In this way, cytoplasmic incompatibility leads to a steady increase in the frequency of *Wolbachia*, often until the entire insect population is *Wolbachia*-positive.

For *Wolbachia* with antiviral function, the result is an insect population dominated by individuals unable to vector the virus. These traits of *Wolbachia* have been applied to protect human populations against the viruses vectored by the yellow fever mosquito, *Aedes aegypti*.

Wolbachia and suppression of dengue transmission by the yellow fever mosquito

The yellow fever mosquito is a first-order threat to public health throughout the tropics and subtropics. It transmits several deadly viruses, including the agents of yellow fever, dengue, zika, and chikungunya; it is well adapted to urban habitats; and it favours humans for the blood meals by which the virus is transmitted. This mosquito does not naturally bear *Wolbachia* but it can be colonized by *Wolbachia* derived from *Drosophila* fruit flies, and the capacity of these *Wolbachia*-harbouring mosquitoes to transmit the dengue virus is reduced by over 1,000-fold under the laboratory conditions.

Building on these laboratory studies, yellow fever mosquitoes infected with *Wolbachia* have been released into the wild. As in the laboratory studies, the *Wolbachia* in these mosquitoes were derived from fruit flies. At first sight, this may appear 'unnatural', but it is not a cause for concern because, although *Wolbachia* is mostly transmitted from the mother insect to her eggs, it also naturally moves around between different insect species.

The first releases were made in 2011 at two residential locations near Cairns in Australia, with strong support from local residents. Within four months, nearly all the yellow fever mosquitoes at these sites had become *Wolbachia*-positive and, when tested, they were unable to transmit dengue virus. Monitoring over subsequent years has confirmed that the mosquito populations at these sites remain predominantly *Wolbachia*-positive and unable to vector the dengue virus. Yellow fever mosquitoes bearing

Wolbachia have now been released in various locations in the tropics, including Indonesia, Fiji, Vietnam, Brazil, and Mexico. Interim results are indicating reduced incidence of dengue virus and disease in the local human populations.

Ongoing research is also investigating the potential of *Wolbachia* to suppress the transmission of other disease agents vectored by insects. These include medically important insect vectors, notably *Anopheles* mosquitoes that vector *Plasmodium* species, eukaryotic microorganisms that cause malaria. There is also increasing interest in applying the technology to insect vectors of disease agents affecting major crops. Notably, proof of principle has been obtained that rice planthoppers bearing *Wolbachia* display much-reduced transmission of the rice ragged stunt virus, which can reduce rice grain yield by 50 per cent or more in South-East Asia.

Scientists are fully aware of various risks associated with this technology. One risk is that the efficacy of the antiviral function and cytoplasmic incompatibility of *Wolbachia* may decline over time, perhaps linked to shifts in temperature or diet. There is also the hypothetical concern that *Wolbachia* may facilitate transmission of other pathogens, including emerging disease agents that we are currently unaware of. Nevertheless, there is great optimism that this new technology, together with other strategies, will dramatically alleviate the burden of many insect-vectored diseases of importance for public health and agriculture.

Chapter 6
Plant microbiomes in agriculture and food production

Plants dominate most terrestrial ecosystems, and they provide most of our food. Today, half the world's habitable land is devoted to food production, much of it by unsustainable methods. Many of the strategies to promote sustainable agricultural practice involve plant microbiomes. Some of the microbial players have been known since the 19th century; these include the mycorrhizal fungi which colonize many plant roots and the rhizobia which inhabit root nodules of legumes. The importance of other plant-associated fungi and bacteria has become widely appreciated only with the advent of advanced sequencing technologies that have also enabled the study of animal microbiomes (see box: **Molecular methods for studying microorganisms** in Chapter 1). The rapidly accumulating knowledge of plant microbiomes is being used to augment soils and plants with microorganisms that promote sustainable crop production.

'Getting in on the ground floor'

The first land plants harboured a microbiome, and this collaboration has played a key role in the success of both plants and their microbial partners. Both fossil and molecular data indicate that the land plants evolved from freshwater green algae about 420 million years ago. In some of the earliest plant fossils, the internal structure is preserved, revealing fungal hyphae ramifying through the tissues of the plant. Some of these fossilized fungi display massively branching, miniature tree-like structures that resemble the arbuscules found today only in the arbuscular mycorrhizal fungi (AM fungi) of plants. The AM fungi are an ancient group, and it is most likely that the arbuscule-forming fungal partners of the first land plants were the ancestors of modern AM fungi.

Why were the first land plants intimately associated with AM fungi? The answer is that the partnership with AM fungi solved one of the biggest problems in the transition from water to land: that nutrients, accessed by algae from the water column, had to be acquired from the terrestrial substrate. The first plant fossils comprised a leafless vertical shoot arising from a horizontal, unbranched structure with hair-like projections known as rhizoids. One group of land plants, the liverworts, which diverged early in the evolutionary history of plants, also have rhizoids but no roots. Liverworts obtain nutrients and water from the substrate via both their rhizoids and AM fungi which colonize the entire plant and extend into the substrate. It is very likely that AM fungi enabled the first land plants to access nutrients in the same way as for liverworts.

Plant roots are evident in the fossil record from 390 million years ago, initially in the earliest woody plants and plants that resemble modern tree ferns. The AM fungal partners were retained but they became restricted to the roots, as for modern plants with roots. Today, an estimated 70 per cent of plant species are associated with AM fungi. The AM fungi have been replaced by different fungi in a minority of plants, notably the ectomycorrhizal fungi of many trees in temperate regions, and a few plants do not associate with mycorrhizal fungi at all (for details see box: **The diversity of mycorrhizal fungi**).

The reason why plant scientists are confident that AM fungi solved the nutrient acquisition problem for the first land plants is

The diversity of mycorrhizal fungi

Mycorrhizal fungi are defined as fungi which are associated with plant roots and contribute to plant health, primarily by promoting nutrient uptake. Some mycorrhizal fungi can also protect the plant against pathogens, toxic metals, and drought stress.

Arbuscular mycorrhizal fungi (AM fungi) probably colonized the first land plants (see text) and they are associated with an estimated 70 per cent of modern land plant species. The AM fungi are the dominant members of an ancient division of fungi, the Glomeromycota. They have been replaced by different types of mycorrhizal fungi in some plant groups, including the following:

Ectomycorrhizal fungi are readily visible to the naked eye as a distinctive sheath around the tips of short, thick roots. The fungus also extends into the roots. Ectomycorrhizal fungi have evolved many times within the last 50 million years, and they include representatives of the Basidiomycota, Ascomycota, and Zygomycota. They primarily colonize trees in temperate and boreal habitats, including many conifers and members of the beech, willow, and birch families. Some trees, for example willows, can bear both ectomycorrhizal and AM fungi.

Orchid mycorrhizal fungi are found associated with all orchids at the seedling stage. Orchid seeds are tiny and contain insufficient nutrients to support the germinating plant. Instead, seedling growth is dependent on sugars and other nutrients provided by their mycorrhizal fungi. Some orchids (e.g. ghost orchids) have little or no capacity to photosynthesize and they are dependent on their mycorrhizal partner for sugars throughout their lifespan.

Ericoid mycorrhizal fungi occur only in the Ericales, which include the heathers and rhododendrons, and these associations are very abundant in acidic heathland soils. Most ericoid mycorrhizal fungi are Ascomycota, and they protect the plant roots against the toxic levels of free metals in these soils, as well as providing nutrients.

Non-mycorrhizal plants, i.e. plant species that do not associate with mycorrhizal fungi, are widespread in several plant families, including the Brassicaceae (cabbages and mustards) and Caryophyllaceae (the pinks and carnations). Non-mycorrhizal plants have alternative strategies for efficient nutrient acquisition, and many species thrive in nutrient-rich soils.

that AM fungi provide modern plants with key nutrients, such as phosphorus. This has been demonstrated many times by a standard experimental procedure. The plants are grown in soil that is deficient in phosphorus or supplemented with phosphorus, either with or without an inoculum of AM fungi. In the supplemented soil, the plants grow vigorously, whether or not they have access to the mycorrhizal fungi; but in the low-phosphorus soil, only the plants associated with AM fungi grow well. Experiments of this general design have also demonstrated the nutritional significance of other types of mycorrhizal fungi to plants. Overall, plants in natural ecosystems are estimated to derive much of their phosphorus and nitrogen requirements from mycorrhizal fungi.

The first land plants were not the first organisms on land. They were preceded by various microorganisms, including bacteria and fungi, that formed extensive microbial crusts on and within terrestrial substrates. It is very likely that bacteria derived from this ancient terrestrial community colonized the first land plants, just as modern plants are associated with a diverse community of beneficial bacteria (as considered later in this chapter), although this is not generally evident in the fossil record. Whether the AM fungi were also acquired as plants came onto land or were partners of the algal ancestor of land plants is uncertain.

How mycorrhizal fungal networks function

So far, we have addressed the relationship between mycorrhizal fungi and individual plants. Let us now broaden our perspective to a patch of vegetation. As described above, the fungal hyphae extend from an individual root into the soil, from where they capture nutrients. The hyphae can reach up to 20 cm from their host plant, giving them the opportunity to contact roots of other plants. Because many mycorrhizal fungi, especially AM fungi, can form associations with a very wide diversity of plant species, a single fungus can form underground connections between different plants, resulting in a complex network of fungal–plant associations.

The underground network of mycorrhizal fungi is more than a physical connection between plants. It influences nutrient acquisition by the mycorrhizal plants. Consider a single mycorrhizal fungus that is connected to two plants. The fungus acquires phosphorus from the soil and delivers it to the roots of its two plant partners. But how much is transferred to each of the two plants? In many situations, the phosphorus is transferred preferentially to the greater phosphorus sink, meaning that the more phosphorus-hungry plant is provided with more phosphorus. Some mycorrhizal fungi appear to function by the maxim 'to each according to need'.

There are, however, complications. Mycorrhizal fungi do not deliver soil nutrients to their plant partners for free. The plants feed the fungi with sugars and lipids that can account for up to 20 per cent of the carbon that a plant obtains by photosynthesis. There is evidence that the fungi can modify the allocation of phosphorus, so that more goes to the plant that provides them with more carbon. Plants can also control the supply of carbon to their mycorrhizal fungi, favouring the fungus that provides the most phosphorus, and in some situations, delivering no carbon to fungi from which they derive little or no phosphorus. In the context of these data, the relationship between mycorrhizal fungi and plants has been described as a biological market, in which the partners trade nutrients.

So, is the mycorrhizal fungal network the great equalizer or a market trader that favours partners with resources? As so often in biological systems and especially interactions involving beneficial microbes, the answer is that it depends. A fruitful approach to investigate this issue quantifies the growth of plants of unequal size, for example a seedling and mature plant, connected by a common mycorrhizal fungus. These studies can be conducted on potted plants in the laboratory or in natural habitats. In many studies, seedling growth is improved by linkage via mycorrhizal fungi to a larger plant, but a minority of studies report depressed growth of linked seedlings or no effect at all. The reasons for these contradictory results are not understood fully, although plant species, type of mycorrhizal fungus, soil type, and water availability have all been implicated.

The variability of the interactions between plants and the underground mycorrhizal network is recognized as an important factor influencing plant diversity in natural ecosystems. It also has major implications for crop production.

Mycorrhizal fungi in crop production

Over the last century, crop production has become dependent on high nutrient inputs to maximize yield, together with aggressive tillage (mechanical disruption of the soil) and chemical pesticides to control weeds, pathogens, and insect pests. These technologies are increasingly recognized as unsustainable, due to soil erosion, nutrient leaching, and resistance to herbicides and pesticides, collectively resulting in environmental degradation and declining crop yields. There is widespread interest in low nutrient input, no-tillage (or low-tillage) agricultural practices that promote crop productivity, and biological control of pests, particularly in the context of polycultures, i.e. growing several crops together. Improved management of the associations between crops and mycorrhizal fungi is becoming an important factor in the design of sustainable agricultural practices.

Most crop plants other than the brassicas are mycorrhizal. For example, the yield of peas and beans, onions, and solanaceous plants, including potatoes, tomatoes, and eggplants, is promoted by AM fungi, especially under low-nutrient conditions, but the yield advantage of AM fungi for the main cereal crops (e.g. wheat, rice, maize) is variable. Some plants, including crops, however, can benefit from the association with mycorrhizal fungi in other ways, including increased competitiveness against weeds and increased resistance to some pathogens and herbivores. These traits facilitate reduced pesticide applications and are especially important for organic farmers.

Two aspects of intensive agriculture are particularly deleterious to AM fungi: high nutrient inputs and tillage. Heavily fertilized agricultural soils suppress the colonization of plant roots by most mycorrhizal fungi, but they can promote non-beneficial fungal strains that consume plant carbon while providing limited nutrients in return. In addition, the mycorrhizal networks are disrupted in soils that are frequently and deeply tilled, further reducing the abundance and diversity of the mycorrhizal fungal communities.

Various studies are demonstrating that changes to agricultural practice, especially restricting fertilizer applications and tillage, together with polyculture plantings, can promote soil communities of AM fungi. A current priority is to develop methods that manage the AM fungi more effectively, to favour species that optimize soil health and crop yield. To reach this goal, some practitioners are augmenting soils with mycorrhizal fungal strains that are available commercially or propagated from local soil by the individual farmer. The desirable fungi can be added to the soil at the time of sowing, or they can be administered as a coating to seeds.

Fixing crop production by microbial nitrogen fixation

Nitrogen is plentiful, with nitrogen gas (N_2) accounting for 78 per cent of the air, but most organisms cannot utilize nitrogen gas to synthesize proteins and other nitrogen-containing molecules needed for growth. Because of this metabolic limitation, the growth of many plants is limited by nitrogen, and nitrogen fertilizers play a central role in commercial crop production.

Exceptionally, some bacteria can convert nitrogen gas into ammonia, a process known as nitrogen fixation. Although no eukaryotes have the genetic capacity to fix nitrogen, various plants gain access to this capability by forming associations with nitrogen-fixing bacteria (Table 2). Two types of association are particularly relevant to crop production. The first is nitrogen-fixing bacteria, generically known as rhizobia, in root nodules of many leguminous plants. Legumes harbouring rhizobia are less dependent on nitrogen fertilizer than most other crops, and they improve soil nitrogen content for other crops when grown in rotation or co-culture. The second type of association involves various nitrogen-fixing bacteria that live on the surface or within the tissues of many plants without inducing nodules or other plant structures visible to the naked eye. The rhizobia in legume root nodules have been estimated to fix 50–465 kg N ha^{-1} yr^{-1}, about four times the rate of other plant-associated nitrogen-fixing bacteria.

Concerns about the sustainability of current agricultural practice have led to heightened interest in plant-associated nitrogen-fixing

Table 2. Associations between plants and nitrogen-fixing bacteria

Nitrogen-fixing bacteria	Plant hosts
Cyanobacteria, e.g. *Nostoc*	Many cycads, *Azolla* water ferns, various liverworts and hornworts
Frankia species (these bacteria are members of the phylum Actinobacteria)	Various trees and shrubs (members of 8 dicot families: Betulaceae, Casuarinaceae, Coriariaceae, Dastiscaceae, Elaeagnaceae, Myricaceae, Rhamnaceae, Rosaceae)
Rhizobia, including species of *Rhizobium*, *Bradyrhizobium*, *Sinorhizobium*	Many legumes, including all annual crops, e.g. peas, beans, and forage crops, e.g. clover, alfalfa
Diverse bacteria, e.g. *Azospirillum*, *Azoarcus*, *Herbaspirillum* (often referred to as 'associative nitrogen fixers' and as examples of plant growth-promoting rhizobacteria (PGPRs), which are defined by their association with plant roots. However, many PGPRs do not fix nitrogen, and some nitrogen-fixing bacteria are associated with the shoots of plants as well as the roots)	Many plants

bacteria. The food supply for approximately half the world's population is dependent on synthetic nitrogen fertilizer produced by the Haber–Bosch reaction. This process converts nitrogen gas into ammonia under conditions of very high temperature and pressure. It consumes 1 tonne of natural gas (or other fossil fuel) per tonne of ammonia produced, and up to half of the fertilizer applied to agricultural fields is lost to the atmosphere or in runoff, causing environmental degradation.

Greater use of nitrogen fixation by plant-associated bacteria would ensure sufficient soil nitrogen for crop growth with less environmental damage than caused by synthetic fertilizer produced by industrial nitrogen fixation. Nevertheless, improved biological nitrogen fixation can only be part of the route to sustainable fertility of agricultural soils because plant growth also requires adequate supplies of other nutrients, including phosphorus, potassium, and various micronutrients.

Elite rhizobia in legumes and other crops

Although rhizobia in legume root nodules can fix nitrogen at very high rates, the agricultural benefits of these associations are constrained by two factors: fixation rates in the field are often meagre, and the most important staple crops for human populations are not legumes but cereals, which do not associate with rhizobia. Much research has been devoted to addressing these problems.

Leguminous plants acquire their complement of rhizobia from the soil. However, some natural populations of rhizobia colonize the plant roots aggressively but fix nitrogen at low rates and fail to promote plant growth. In addition, the populations of infective rhizobia can be very low in some soils, especially agricultural soils where non-legume monoculture crops have been grown for many years.

In principle, these difficulties can be ameliorated by inoculating rhizobia administered at the time of sowing. Various commercial inoculants are available, either as a seed coating or soil additive. They are often referred to as elite strains because they can both fix nitrogen at high rates and colonize many different leguminous crop plants. However, their efficacy in field conditions is variable because of poor tolerance of some soil conditions and inability to compete with indigenous rhizobia and other soil bacteria. There is much ongoing effort to produce inoculants optimized for different conditions, including soils of different types and containing different profiles of competing microorganisms.

A far more ambitious strategy to expand rhizobial nitrogen fixation in agriculture is to create associations with non-legumes, especially cereals. If this scientific breakthrough is ever achieved, it would dramatically reduce the input of nitrogen fertilizers.

In experimental studies, rhizobia administered to rice, wheat, or barley plants often colonize the root surface and may even penetrate via cracks to internal tissues. Chemical treatments that partially degrade the root cell walls of rice plants prior to exposure to rhizobia facilitate rhizobial entry and the formation of nodule-like structures, but the rhizobia do not attain high densities or fix nitrogen at appreciable rates. An alternative and likely more productive approach is genetic modification of cereal plants, so that they express genes required for nodule formation. This goal is potentially feasible in the context of evidence that the same plant signalling, the Sym pathway (see Chapter 2 and Figure 3), contributes to root colonization by rhizobia and AM fungi. Cereal crops possess the Sym pathway and their roots are colonized by AM fungi. Nevertheless, much painstaking research is needed for cereals with nitrogen-fixing rhizobia to become a reality.

A cornucopia of nitrogen-fixing bacteria in plants

Recent years have witnessed an increasing interest in nitrogen-fixing bacteria other than rhizobia as a promising route to enhance the nitrogen nutrition of non-legume crops. This approach has been stimulated by unexpected discoveries of nitrogen-fixing bacteria associated with crops and their relatives, especially in the tropics.

One discovery is that the stems of sugarcane varieties in Brazil often bear communities of nitrogen-fixing bacteria that contribute 10–30 per cent of the plant's nitrogen. A strain of one of these bacteria, *Gluconacetobacter diazotrophicus*, has been isolated into culture. Under experimental conditions, this bacterium has been reported to colonize other crops, including rice, maize, wheat, potato, and tomato, improving crop nitrogen content and yield.

A further intriguing discovery is that some traditional varieties of maize in Mexico have many aerial roots that release mucilage, in which nitrogen-fixing bacteria are trapped. These bacteria release much of the fixed nitrogen as ammonia, which is taken up by the plant. Under conditions of high rainfall, which promotes mucilage production by the aerial roots, this association can provide up to 80 per cent of the plant's nitrogen requirement. Transferring this association to commercial maize cultivars poses some tough but biologically fascinating problems, both to promote aerial root production (a trait that has been lost in breeding programmes for modern maize varieties) and to facilitate sustained nitrogen fixation independent of rainfall.

There is every expectation—and some evidence—that other traditional crop varieties and other plant species are associated with a range of nitrogen-fixing bacteria that can potentially be harnessed for improved crop production. One focus of interest is *Azospirillum*, bacteria that are borne on the surface and internal tissues of many different plants, where they fix nitrogen. There is also the tantalizing possibility of re-creating legume-like root nodules in cereals, but with *Azospirillum*, not rhizobia. Specifically, the roots of rice plants can be induced by plant hormones to produce swellings, known as paranodules, and *Azospirillum* colonize these paranodules to high density, where they fix nitrogen at elevated rates.

Various species and strains of *Azospirillum* are commercially available in diverse countries for use as soil inoculum to enhance crop yield, especially of soybean, wheat, and maize. Although some scientific trials have reported strong stimulation of plant growth and crop yield, the efficacy of these inocula can be variable.

There are also indications that the observed stimulation of plant growth is often unrelated to nitrogen fixation rates and can be attributed to other types of interaction between *Azospirillum* and the plant, as discussed below.

In summary, approaches to harness nitrogen-fixing bacteria for sustainable crop production have yielded mixed results. While much is being achieved with the development and application of protocols for improved nitrogen fixation by crop-associated bacteria, some goals (e.g. nodulation of cereals) have yet to be achieved. Some scientists have responded to this situation by considering a complementary strategy. Instead of identifying a valued microbial function (e.g. nitrogen fixation) and investigating how it can be optimized for crop growth, they have studied plant-associated microorganisms that promote crop growth, irrespective of the nature of their interactions with plants. The main focus of this second strategy is plant-associated microorganisms that are not evident as conspicuous structures. The remainder of this chapter is devoted to these associations.

Plant growth-promoting rhizobacteria: the PGPRs

Plants germinated from surface-sterilized seeds in microbe-free substrates generally grow less well and are more susceptible to pathogens than plants raised without these microbe-excluding precautions. These findings are strongly reminiscent of the findings for mice and other animals raised under microbe-free conditions, as discussed in previous chapters. They are entirely to be expected because both animals and plants have evolved and diversified in the context of long-standing interactions with microorganisms (see Chapter 1).

Most of the microorganisms recovered from untreated plants are root-associated bacteria, also known as rhizobacteria. These bacteria have been studied intensively, and individual strains and small communities of bacteria that promote plant growth are widely known as PGPRs (plant growth-promoting rhizobacteria).

Two attributes of PGPRs play a large role in supporting enhanced plant growth. One is that they make soil nutrients more readily available to plants. The underlying mechanisms include the capacity to solubilize soil phosphate, to fix nitrogen, and to convert complex organic nitrogen sources to soluble forms, especially ammonia and nitrate, which are readily taken up by the roots of many plant species. The nutrient-provisioning functions of PGPRs can be particularly important for plant species, such as brassicas, which do not form associations with mycorrhizal fungi. For plant species that are mycorrhizal, PGPRs can sometimes function in tandem with mycorrhizal fungi, such that nutrients released by PGPRs are taken up by mycorrhizal fungi, from which they are delivered to plants. In other situations, plants may gain access to nutrients generated by PGPRs only after the bacterial cells die.

The second way in which PGPRs promote nutrient acquisition by roots is by their stimulation of root growth. Rapid root growth of newly germinated seedlings provides secure anchorage and access to nutrients and water, and a more extensive root system of established plants enables larger volumes of soil to be explored for nutrients and water. This effect is mediated by the capacity of many PGPRs to synthesize plant hormones, especially auxins. Plant scientists have long known that auxins produced in the plant stem and transported to the roots promote the production and sustained growth of roots. It appears that this plant function is augmented by the supply of auxins from bacteria in the rhizosphere.

Why do bacteria associated with plant roots release the auxin plant hormones? Auxins have no known function in the bacteria, and they are only produced by bacteria that are associated with plants. This suggests that the bacterial auxins are produced to influence plant biology in ways that favour the bacteria. (As considered in Chapter 5, in relation to resident microorganisms of animals, we should expect rhizobacteria to function out of self-interest.) The stimulation of root growth by bacterial auxins may provide more root-associated habitat for the auxin-releasing bacterial cells and their progeny. There is also evidence that the auxins promote the release of nutrient-rich root exudates that support bacterial proliferation in the rhizosphere. In other words, the small cost of auxin synthesis by the bacteria is amply compensated by the benefits of associating with a growing, exudate-releasing plant partner.

PGPRs have also been implicated in improved plant tolerance of soils with low water content. Research on *Arabidopsis* plants has shown that PGPRs influence the internal architecture of roots, specifically the structural organization of a layer of cells known as the endodermis. The cell walls of the endodermal cells are modified to impede the movement of water between the innermost tissues and outer regions of the root. PGPRs make the root endodermis more permeable by modifying hormonal signalling in the root, resulting in reduced deposition of a lipid-like substance called suberin into the endodermal cell walls. The effects of these structural changes extend beyond improved water relations of the plant to enhanced tolerance of low availability of soil nutrients, as well as high salt levels caused, for example, by long-term irrigation of soils.

Overlaying their beneficial effects on the water and nutrient relations of plants, some PGPRs protect plants against pathogens. The underlying mechanisms have many parallels to the mechanisms of colonization resistance by animal microbiomes (Figure 9). The root-associated bacteria variously produce toxins that kill or inhibit pathogens, outcompete pathogens for nutrients and space, and induce plant defences that are more damaging to pathogens than to themselves. These traits play a central role in the disease-suppressive soils discussed in Chapter 5. Bacteria associated with plant shoots, including leaves and flowers, play a similar defensive role.

PGPRs and crop production

There is much enthusiasm to harness the beneficial effects of rhizosphere bacteria on plants for improved crop production. However, the rhizosphere microbiome is diverse and highly variable in composition, and its influence on plant traits varies with these microbial factors, with plant species, genotype, and age, and with environmental conditions. Plant scientists and microbiologists are working to make the effects of the rhizosphere microbiome on plant traits more reliable and robust by developing beneficial microbes for use as soil inoculants. The main focus is bacteria that are readily culturable and can be packaged and delivered as soil inocula. One widely adopted example already introduced in this chapter is the various isolates of *Azospirillum* bacteria that can fix nitrogen (see above), promote soil nutrient uptake, and protect against root pathogens. However, as mentioned previously, the efficacy of *Azospirillum* inoculants can be variable and sometimes they are totally ineffective.

One approach to reduce the variability of PGPR effects on crop production is to design small communities of microorganisms with overlapping functions, with the expectation that one or more of the microbes will be beneficial under different circumstances. These small 'synthetic' communities comprise up to 20 microorganisms, usually bacteria that have been isolated from the rhizosphere of the plant of interest. Computational and experimental studies assist in the selection of bacteria for these synthetic communities. For example, sophisticated computer modelling can predict how environmental factors, including nutrient supply, can influence the exchange of nutrients among the bacteria and nutrient release from the community to plant. These predictions can then be tested using microbes-on-a-chip technologies made possible by recent advances in microfabrication and microfluidics. Briefly, thousands of microbial communities grow in tiny wells on a slide, nourished by a medium of defined composition that flows across the slide, enabling the performance of each community to be tested under each of many different nutrient conditions. A subset of these experimental communities can then be tested for efficacy in plants. In some studies, high-performing communities are then taken through several generations of plants, and communities that display a progressive improvement in the desired traits (e.g. supporting improved plant growth or tolerance of low water availability) are selected. These studies are currently in the experimental stage, but they offer great promise for improved reliability of PGPR soil inoculants over the coming years.

Endophytic fungi

The research on PGPRs is increasingly being complemented by studies of plant-associated fungi. The fungal communities include the mycorrhizal fungi, considered in the opening sections of this chapter, and other fungi that live entirely within the plant tissues, often ramifying extensively between plant cells in the shoot system or roots. These fungi are commonly known as endophytes.

By far the best-studied fungal endophytes live in grasses. Many of these fungi produce alkaloids that are toxic to grazing animals. The primary target of the alkaloids is the animal nervous system, and livestock affected by these toxins exhibit symptoms graphically described as 'the jitters' or 'staggers'. This fungus-mediated defence of grasses against herbivory has been eliminated from most improved pastures used in livestock production.

Many plants other than grasses also bear fungal endophytes but, unlike the grass endophytes, these fungi are widely described as 'symptomless', meaning that they are neither detectably beneficial nor harmful to the plant. Many of these fungi may be biding their time, waiting for the leaf or plant to die, when they will be the first to derive nutrients as the dead plant tissue decomposes. However, their impact on the plant can vary widely with environmental conditions, the genetic make-up of the plant, and the activities of co-occurring microorganisms. Endophytic fungi of the same species and sometimes even genetically identical isolates can act either as pathogens, especially in plants with a weakened defence system, or as beneficials that defend the plant by competing with co-occurring potential pathogens.

Further insight into fungal endophytes has come from studies of plants living in extreme habitats. For example, a survey of plants in the vicinity of hot geothermal soils of Yellowstone National Park (Wyoming, USA) revealed that some plants were thriving in soils at temperatures up to 60°C. When the endophytic fungi from these plants were transferred into plants that are intolerant of high temperatures, such as tomato or watermelon, these latter plants thrived at elevated temperatures. Other endophytic fungi have been isolated from plants growing under extremely cold or saline conditions or in metal-contaminated soils and, as for heat stress, they include culturable strains that protect a variety of crops against these environmental stresses. Cocktails of these fungi are now commercially available in various formulations, including products that also contain PGPRs. These products can yield improved plant growth and pathogen resistance for various crops, including maize, cotton, soybean, and wheat.

The promise of plant microbiomes in agriculture

Plant microbiomes are already making an important contribution to more sustainable agricultural systems. The abundance of evidence that plant-associated microorganisms contribute to plant growth and protection against pathogens and environmental stresses provides the basis for strategies that complement or replace synthetic fertilizers and chemical pesticides. Some strategies to inoculate soils or seeds with beneficial bacteria or fungi have already been developed as commercial products. The scope of these products varies. Many can be used for a range of crops in many locations, while other products involve microorganisms which are tailored to specific locations, soil types, crops, etc. A high research priority is to understand better how microbes benefit plants, and to identify optimal microbial strains and agricultural practices to improve the reliability of microbiome-based strategies for crop production. As we will see in the next and final chapter of this book, many of these opportunities and difficulties are shared by the practitioners developing microbial therapies for improved human health.

Chapter 7
Microbial therapies and healthy microbiomes

The human microbiome holds great promise for new approaches to prevent, ameliorate, and cure disease. Biomedical research effort is framed largely by the concept of dysbiosis, a microbiome that causes or exacerbates disease, and it has two complementary goals. The first is to shift the microbiome in sick people from a dysbiotic state to a health-promoting homeostatic state, with the expectation that this change will lead to improved health. The second goal is to support the homeostatic microbiome in healthy people, predicting that this will be protective against disease.

This final chapter will describe some of the strategies to promote human health by modification of the gut microbiome, summarized in Figure 12. Some approaches are already in the clinic and saving lives; others are still at early stages of laboratory investigation. Much of this research is driven by a concern that the human microbiome is becoming impoverished by aspects of the industrial lifestyle, typified by a diet of ultra-processed foods, extensive use of antimicrobials, and lack of exercise, with the implication that microbiome interventions are acts of restoration, rather than modification.

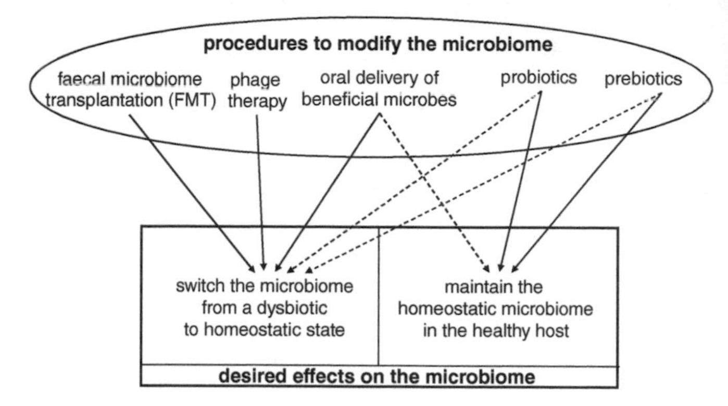

12. Gut microbiome modification for improved human health. Five procedures (top) can be used to either drive the microbiome from a disease-promoting (dysbiotic) to health-promoting (homeostatic) state, or to maintain the homeostatic state of the microbiome. See Chapter 1 and Figure 2 for further explanation of the dysbiotic and homeostatic microbiome. Probiotic refers to living microorganisms in food that promote good health by supporting a beneficial microbiome; prebiotic refers to a food item that supports beneficial microorganisms in the gut. Solid arrows and broken arrows refer to the principal and minor application of the procedure, respectively.

The effect of lifestyle on the microbiome

It is argued by some microbiome scientists that the microbiome of many people is being perturbed by industrial lifestyles. Most of the available data relate to the gut microbiome, but similar concerns have been raised in relation to the skin microbiome.

Disquiet about the effects of lifestyle on the human gut microbiome has been fuelled by studies of different human populations. The gut microbiome of people with the industrial lifestyle is different and less diverse than both traditional agricultural populations and hunter-gatherer communities such as the Hadza of Tanzania and Yanomami Amerindians in Brazil and Venezuela. These data have been interpreted as evidence that the industrial lifestyle has led to the loss of many microbial species. Consistent with this perspective, the microbiome of 1,000–2,000-year-old palaeofaeces samples is more similar in composition and diversity to modern rural populations than to industrial populations. A key factor implicated in the depletion of the gut microbiome is the typical industrial diet, which has little fibre and is rich in fats, sugars, and meat, compounded by the inclusion in ultra-processed foods of emulsifiers that suppress many gut microbes. In addition, extensive medical use of antibiotics diminishes the microbiome, while antimicrobial household cleaning agents restrict the opportunities for transmission of microbes between people.

It is not surprising that the gut microbiome differs between humans with very different lifestyles. However, microbiomes of reduced diversity might be indicative of impaired function. In diverse microbial communities, many microbial species mediate essentially the same function, such as the degradation of dietary fibre in the human colon, and the function is largely unaffected if a few species are lost. But if the overall diversity is low, then function can be severely compromised when a single species is lost. Some critical functions, especially those involved in regulation of immune function, are mediated by species that are at moderate or low abundance, and these functions are particularly vulnerable to processes that reduce diversity.

A set of experiments conducted on laboratory mice illustrates how an inappropriate diet can permanently reduce microbiome diversity. Germ-free mice were colonized with faecal microorganisms from healthy humans, and then provided with either a mixed fibre-rich diet of grains and leafy foods or a low-fibre diet of sugar and cornmeal. The diversity of the gut microbiome was reduced in the mice on the low-fibre diet, relative to the control mice on the high-fibre diet, and the difference increased across the four generations of the experiment. Importantly, when mice were transferred from the low-fibre to the high-fibre diet, their gut microbiome did not recover to the diversity in the control mice, and the failure of recovery became more pronounced in the later generations of the experiment. In other words, an inappropriate diet consumed over several generations can lead to irreversible and progressive loss of microbes. We should, however, treat this experiment as proof-of-concept. Results obtained for inbred mice in laboratory cages cannot be extrapolated to conclude that the microbiome in people with an industrial lifestyle is declining, generation by generation.

Nevertheless, it appears that humans and laboratory mice are not the only species for which inappropriate diet is linked to changes in the gut microbiome. There is also evidence for depletion of the microbiomes of wild animals in habitats dominated by humans. Urban populations of house sparrows, rats, and coyotes, all of which consume ultra-processed food provided or discarded by humans, have a different and less diverse gut microbiome than populations in rural areas. Similar issues arise for animals in zoos, where uniform diet, regular contact with humans, and veterinary use of antibiotics contribute to microbiomes that are species-poor and less distinctive than wild animals of the same species. Some aspects of microbiome loss have been implicated, albeit not conclusively, in the poor health and low fertility of some captive animals. For example, the susceptibility of captive cheetahs to bacterial infections may be linked to the impaired colonization resistance of their perturbed gut microbiome, and ill-health in captive koalas may be related to the depletion of gut microbes that degrade tannins in their tannin-rich diet of eucalyptus leaves.

Hygiene and the impoverished microbiome

The demonstration in the 19th century that bacteria can be agents of disease was transformational for public health. Medical protocols for extreme cleanliness became a priority to prevent disease, and the wider community adopted higher standards of personal cleanliness. These behavioural changes, together with antibiotics introduced in the mid-20th century, led to dramatic decline in infectious disease in developed countries. In parallel with this improving situation for infectious disease, there has also been a dramatic increase in the incidence of childhood atopic disease, i.e. an exaggerated immune response causing asthma, hay fever, eczema, etc.

The first claim of epidemiological evidence that excessive cleanliness may be linked to atopic disease came with the finding that the incidence of atopic disease is greatest in first-born children and children of small families. It was proposed that unhygienic contact with older siblings brought children into regular contact with environmental microorganisms and microbes causing minor illnesses, and that these challenges educated the developing immune system, conferring protection against atopy. Consistent with this interpretation, some follow-on studies found that atopy is also significantly reduced in children growing up on farms or with household pets.

In recent years, the hypothesized link between hygiene and atopic disease has been broadened in two ways. The first is to include other immune disorders that have also increased dramatically in recent decades: food allergies, autoimmune diseases, and the chronic inflammation that is associated with metabolic disease and obesity. The second is to include deprivation of microorganisms contributing to the gut and skin microbiome, as well as a deficit of environmental allergens and minor pathogens, as candidate factors promoting immune-related disease. It is increasingly argued that the healthy development of the immune system depends not only on exposure to 'dirt' and common childhood infections, but also on access to the diversity of beneficial microbes that thrive only in our bodies.

These ideas are plausible because immunological studies demonstrate the importance of microbes, especially the microbiome of the gut and skin, in the healthy development of the immune system (see Chapter 5). However, other lines of evidence are not fully supportive. In particular, there is increasing evidence that the relationship between birth order and incidence of atopy is best explained by pregnancy-order differences in the maternal immune system; poor diet and smoking, not birth order, are the strongest predictors of childhood atopy. Some issues remain confused. For example, different studies on the relationship between self-reported household cleanliness and atopy have yielded positive, negative, and null effects.

Possible long-term health effects of antibiotics administered to infants have also received detailed scrutiny. Several epidemiological studies have revealed that antibiotics (especially multiple treatments) in children under 24 months are statistically associated with childhood eczema and hay fever, as well as above-average body weight, in later childhood. It is tempting to link these effects to antibiotic-induced perturbation of the microbiome, but we should not forget that the infection which triggered the antibiotic treatment may also contribute to these long-term health consequences.

In summary, there are clear indications from experimental research on immune system function in animals and epidemiological studies on human populations that sustained good health is promoted by exposure to microorganisms, especially microbes that contribute to the resident populations inhabiting the gut and skin. The extent to which cleanliness and antibiotic treatments may have reduced microbiome diversity, and thereby contributed to the increased incidence of atopy and other diseases, is far less certain. Nevertheless, there is considerable interest in strategies to maintain and restore a health-supporting microbiome.

Faecal microbiome transplantation by colonoscopy

Clostridium difficile is a serious human pathogen that resides in the colon. This bacterium occurs at low abundance in some healthy people, but it can become abundant following antibiotic treatment or in immune-compromised patients, causing symptoms ranging from diarrhoea to severe intestinal inflammation and sepsis. Faecal microbiome transplantation (FMT) from a healthy donor has become the procedure of last resort for control of *C. difficile* infections, and it often effects a complete cure. This procedure involves the delivery of a microbial slurry obtained from donor faeces into the colon of the patient, usually by colonoscopy. It is now known that FMT delivers specific bacteria, including *C. scindens*, that metabolize bile acids in the colon to products that suppress *C. difficile*.

The suppression of *C. difficile* infections by microorganisms delivered from a healthy individual to the patient by FMT can be considered as a strategy to repair a microbiome that is dysbiotic, driving it to the homeostatic stage (see Figures 2 and 12). Ongoing clinical trials are investigating the potential of FMT in alleviating a range of conditions associated with microbiome dysfunction, including inflammatory bowel disease and neurological disorders, such as autism spectrum disorder. To date, these trial results have been much more variable and modest than the success against *C. difficile* infections.

The value of FMT to resolve microbiome-related disorders has to be weighed against the risks of this procedure. A minority of patients have acquired multi-drug resistant bacteria by FMT, and one case study reports that FMT from an overweight donor led to obesity of the recipient that could not be resolved by dietary or exercise interventions (but the FMT procedure did cure a life-threatening *C. difficile* infection of the recipient).

The risks associated with the inadvertent introduction of deleterious microorganisms by FMT using donor faecal samples are recognized fully by the medical profession, and all donor samples are tested extensively for deleterious microorganisms and viruses. This is an important reason why FMT should never be conducted without full medical supervision. An ongoing research priority is to design standardized therapeutic microbial strains to reduce the risks of FMT. The choice of these optimized strains may need to be tailored to the genetic make-up and microbiome composition of the recipient.

Microbes taken by mouth and probiotic foods

Compared to FMT by colonoscopy, oral delivery of beneficial microbes offers a much more convenient and safer route to modify the gut microbiome in patients with dysbiosis. Much current research is focused on approaches to optimize how microorganisms taken by mouth survive and colonize the gut. Nevertheless, this is, in many ways, an old technology, as the Nissle 1917 strain of the gut bacterium *Escherichia coli* illustrates. This bacterium was isolated by a German medical microbiologist, Alfred Nissle, from the faeces of a soldier in the First World War who, unlike his comrades, never contracted infectious diarrhoea caused by the pathogen *Shigella*. When administered to other soldiers, Nissle's isolate protected them from gastro-intestinal infections. By the early 1920s, the bacterium was made commercially available in easy-to-swallow gelatin capsules, under the tradename Mutaflor®. It continues to be produced in Germany, with sales in Europe, Asia, and Canada.

Taking beneficial microbes by mouth need not be high tech. Many people routinely eat foods, for instance yogurt or fermented vegetables, that contain cultures of live microorganisms claimed to promote health, especially of the gut and immune system. These microorganisms are known as probiotics (Figure 12). The microbial content of some fermented foods is undefined, but various commercial probiotic foods use designated bacterial supplements, such as *Lactobacillus* and *Bifidobacterium* species which remain viable during food storage and are safe for healthy consumers. However, cases of infections by these probiotic bacteria in immune-compromised patients have been reported. Clinical trials addressing the health benefits of probiotic foods have yielded very mixed results. Microbiological research is identifying new strains, including microorganisms derived from humans, to improve probiotic efficacy and safety.

There is also considerable interest in genetically modifying probiotic bacteria, so that they deliver health-promoting products in the gut. These products could include drugs, for example antiviral proteins to combat viral infections, anti-inflammatory proteins to alleviate inflammatory bowel disorder, and anti-cancer drugs to destroy tumours. Animal studies have provided very encouraging results, but the transfer of these procedures to the clinic faces major barriers, in relation to both patient safety and the spread of the modified microorganisms to the environment—and to other people. Stringent protocols to contain the modified microorganisms and to control their capacity to produce the drugs will need to be designed.

Phage therapy

Another approach drawing interest is to harness phage to eliminate pathogenic bacteria and suppress dominant bacterial species in dysbiotic microbiomes. Most phages attack a very restricted range of bacteria, providing a great advantage over antibiotics which are generally deleterious to a wide range of microorganisms. However, as for antibiotics, phage treatment can lead to the evolution of resistance in the target bacterial populations.

Phages that attack gut pathogens can be delivered orally, and successful phage therapy has been reported in both animal models and clinical trials. The target of one study was strains of the bacterium *Escherichia coli* that adhere to the wall of the colon, causing inflammation; the administration of phage that attacks these *E. coli* strains led to reduced symptoms in patients with inflammatory bowel disorder. In a different study, phage therapy was used against strains of the gut bacterium *Enterococcus faecalis* which release a cytolytic (cell-destroying) toxin that passes to the liver causing liver damage. The resultant decline in *E. faecalis* ameliorated the liver disease.

As with probiotic bacteria described in the preceding section, various phages are being modified for highly targeted delivery of drugs. One line of research concerns the bacterium *Fusobacterium nucleatum*. There is increasing evidence that this bacterium is carcinogenic; it promotes the proliferation of colorectal cancer cells and it is abundantly associated with these cancerous cells. A phage that specifically targets *F. nucleatum* has been modified to carry chemotherapeutic drugs. When the modified phage was administered to laboratory mice with colorectal tumours in conjunction with standard chemotherapy, the abundance of *F. nucleatum* declined and the tumours progressed significantly more slowly than with standard chemotherapy alone. Modified phages that target other diseases are being developed, with great potential for clinical application.

Prebiotics to feed the microbes

The term prebiotic refers to foods that promote the abundance and activity of beneficial gut microorganisms (Figure 12). The most remarkable natural prebiotic is human milk which, as explained in Chapter 2, includes sugars that cannot be utilized by the baby and function exclusively to support beneficial bacteria in the infant gut.

Various prebiotic supplements are commercially available for use by adults. Generally, they are sources of dietary fibre that cannot be degraded by human digestive enzymes. These foods pass through the stomach and small intestine to the colon, where they are fermented by gut microorganisms, producing fermentation products, such as butyric acid, that promote good health. Natural foods with a high fibre content are also sometimes referred to as prebiotics.

An important development in the design of prebiotic supplements to support beneficial gut microorganisms has come from studies on undernourished children. One study involved a microbiome-informed dietary intervention applied to correct impaired growth of young children suffering from environmental enteropathy, a condition that causes intestinal inflammation and malnutrition. As described in Chapter 3, research on a poor community in Dhaka, Bangladesh, revealed that the gut microbiome differed in composition between malnourished young children and their healthy peers. (All the malnourished children in this study were identified as having 'moderate acute malnutrition' by the criteria of the World Health Organization.)

Based on detailed studies of the differences in the microbiome between healthy and malnourished children, the researchers generated a panel of candidate foods designed to restore a healthy microbiome in the affected children. The foods contained products that were readily available to the mothers (e.g. chickpea flour, rice, lentil, banana) and were all sourced locally in Bangladesh. These candidate foods were first tested in mice and then piglets (the digestive system of a human is more similar to that of a pig than that of a mouse). In these experiments, the animals were colonized with gut microorganisms from the malnourished children, and the efficacy of the complementary foods in improving the microbiome composition and growth of the animals was determined. Three of the food mixtures yielded excellent results in the animal trials.

The next step was to assess the efficacy of the supplementary foods on malnourished children. Following extensive consultation with mothers, 12- to 18-month-old children were allocated to one of four supplementary feeding treatments for four months, the three formulations that restored a healthy microbiome in the mouse and piglet studies and a widely used nutritional supplement for malnourished children, known as RUSF (Ready to Use Supplementary Food). The trial was double-blinded, meaning that neither the mothers nor the medical staff conducting the trial knew which children received the different treatments. The children administered one of the three supplements designed to promote a healthy gut microbiome grew faster and had improved immune system function, bone development, and nervous system function relative to the other treatments, and the microbiome in faecal samples of these children was enriched in species associated with healthy children in the local community.

This study demonstrates that an integrated understanding of the microbiome and nutritional health of children can be used to design locally sourced prebiotic food supplements for malnourished young children. Protecting and, where necessary, restoring a healthy microbiome offers a new and important route to support the growth and development of young children in disadvantaged communities around the world.

Microbiome-inspired pharmaceuticals

Despite the success of some live microbial therapies and prebiotic supplements, as described above, many researchers are very aware of the limitations of these approaches, including poor reproducibility and safety concerns. An alternative strategy is to apply knowledge of the health-promoting products of the microbiome to develop new pharmaceuticals that can be delivered topically or by mouth. For example, the suppression of the pathogen *Clostridium difficile* infections by bacteria delivered by FMT (see above) has been attributed to the bacterial production of secondary bile acids that inhibit *C. difficile*. In principle, the risky and uncertain procedure of FMT could be replaced by administration of the appropriate secondary bile acids. A similar approach may achieve the modified bile acid profiles associated with improved insulin tolerance following bariatric surgery

(Chapter 3 and Figure 6). Small molecules have been implicated in microbiome-associated alleviation of some disease symptoms, for example taurine and 5-amino valeric acid in a mouse model of autism (Chapter 4), providing the basis for novel pharmaceutical interventions for this disorder.

The rationale for microbiome-inspired pharmaceuticals is that these products will interact directly with the human host. A key concern to resolve is the response of the microbiome to the administered compounds, especially to ensure that the microbiome does not inactivate them or even transform them into toxic metabolites. Ongoing research is addressing these important issues.

Prospects for microbiome science

As Albert Einstein has commented, science favours explanations that are as simple as possible, but no simpler. Mainstream biology has, until recently, explained living organisms too simply. The discipline of microbiology largely ignored the complex social life of microorganisms, while the study of animals and plants disregarded their complex interactions with their resident communities of highly social microbes. Assisted by new technologies that facilitate study of microbiomes, biological explanations have become more complicated—and much more interesting.

The take-home message of this short book is that resident microorganisms are crucial to the health and wellbeing of their animal and plant hosts, and that differences in the composition and activities of the microbiome can have important consequences for the function of the host. For humans, the detail of our partnership with our microbiome can shape our metabolic and immunological health, our mental wellbeing, and our susceptibility to infectious diseases. Similar interactions underlie the wellbeing of other animals, and plant-associated microbiomes influence both natural vegetation and crop production, on which we depend.

Inspired by this new-found understanding of microbiomes, scientists have generated many creative routes for application. Improved management of plant microbiomes is contributing to the development of agricultural practices that are both highly productive and sustainable, as needed to feed the expanding human population in the coming decades. Many of the same principles are starting to be applied to promote human health and to ameliorate and, in some cases, to cure disease. On its own, microbiome science cannot solve all our ills. But fully integrated with other biological disciplines, the study of microbiomes offers a more complete explanation of the natural world and exceptional opportunity for public good.

Glossary

Archaea one of the two domains of bacteria. See also *eukaryotes*, *bacteria*, and *Bacteria*.

bacteria organisms with a simple cellular organization in which the genetic material (DNA) is not separated from the rest of the cell contents by an internal membrane. These organisms are sometimes referred to as prokaryotes. See also *eukaryotes*, *Archaea*, and *Bacteria*.

Bacteria one of the two domains of bacteria. Note that Bacteria (capital B) refers to this domain of life, while bacteria (lower-case b) refers to a grade of organization. See also *eukaryotes*, *Archaea*, and *bacteria*.

bacteriophage viruses that attack bacteria. The term bacteriophage is routinely abbreviated to phage.

bile acids complex organic acids derived from cholesterol that contribute to digestion. The primary bile acids are produced in the liver and delivered to the small intestine. Primary bile acids that transit to the colon are transformed by the microbiome in the colon to secondary bile acids.

colon the final region of the gut in humans and other mammals; also known as the large intestine.

colonization resistance suppression of pathogens by the microbiome, thereby protecting the host from pathogen-mediated disease.

dysbiosis a microbiome that causes or exacerbates disease of its host.

endosphere microbiome the microbiome associated with the internal tissues of plants. See also *phyllosphere microbiome* and *rhizosphere microbiome.*

epidemiology study of the patterns of disease in humans. This discipline plays a critical role in shaping policy on public health.

eukaryote organisms in which the genetic material (DNA) is separated from other parts of the cell within a membrane-bound nucleus. Eukaryotes include the animals and plants. See also *Archaea, bacteria,* and *Bacteria.*

fermentation products small organic molecules produced by microorganisms growing under oxygen-free conditions. Microbial fermentation products include acetic acid and butyric acid. See also *respiration.*

germ-free laboratory animals raised from birth under totally microbe-free conditions. Germ-free animals are also referred to as axenic or gnotobiotic.

host the plant or animal that harbours a microbiome.

hypha (plur. hyphae) long filament(s) of cells; growth form of many Fungi and Actinobacteria.

metabolism the chemical reactions in living organisms that provide energy and support growth.

microbial therapies procedures that modify the microbiome, leading to improved health of the host.

microbiome the community of microorganisms associated with an animal or plant.

microorganisms organisms that live as single cells for all or most of their lives. Archaea, Bacteria, and many eukaryotes are microorganisms.

neuron nerve cell; the cell type in the nervous system that transmits information to other cells, especially other neurons and muscle.

phage see *bacteriophage.*

photosynthesis the light-dependent fixation of carbon from carbon dioxide into sugars. Plants, eukaryotic algae, and some bacteria display photosynthesis.

phyllosphere microbiome the microbiome associated with the surface of plant shoots. See also *endosphere microbiome* and *rhizosphere microbiome.*

prebiotic a food item that supports beneficial microorganisms in the gut. Most prebiotics cannot be degraded by human digestive enzymes and their utilization by gut microorganisms promotes the production of health-promoting fermentation products.

probiotic living microorganisms in foods that promote good health, generally by supporting a beneficial gut microbiome.

respiration the breakdown of sugars to generate energy. Aerobic respiration uses oxygen and generates more energy than the oxygen-independent anaerobic respiration. See also *fermentation products*.

Rhizobia bacteria that colonize swellings, known as nodules, on the roots of many leguminous plants (e.g. peas and beans).

rhizosphere microbiome the microbiome associated with the surface of plant roots and nearby soil. See also *endosphere microbiome* and *phyllosphere microbiome*.

virus parasites of the cells of bacteria and eukaryotes. The viral complement of an animal or plant is known as the virome. Viruses that attack bacteria are known as bacteriophage.

Further reading

General texts

Antwis, R. E., Harrison, X. A., and Cox, M. J. (eds) (2020). *Microbiomes of Soils, Plants and Animals* (Cambridge: Cambridge University Press).

Blaser, M. J. (2014). *The Missing Microbes* (New York: Henry Holt & Company).

Douglas, A. E. (2018). *Fundamentals of Microbiome Science* (Princeton: Princeton University Press).

Kashyap, P. L., Srivastava, A. K., and Srivastava, M. (eds) (2021). *The Plant Microbiome in Sustainable Agriculture* (Hoboken, NJ: Wiley-Blackwell).

Montgomery, D. R. and Bikle, A. (2016). *The Hidden Half of Life: The Microbial Roots of Life and Health* (New York: W. W. Norton & Company).

Yong, E. (2016). *I Contain Multitudes: The Microbes Within Us and a Grander View of Life* (New York: Harper Collins).

Additional references relevant to individual chapters

Chapter 1: Living with microbes

Blaser, M. J. (1999). The changing relationships of *Helicobacter pylori* and humans: implications for health and disease. *The Journal of Infectious Diseases* 179, 1523–30.

Douglas, A. E. (2018). Which experimental systems should we use for human microbiome science? *PLoS Biology* 16, e2005245.

Eisenstein, M. (2020). The hunt for the healthy microbiome. *Nature* 577, 56–8.
Hooks, K. B. and O'Malley, M. A. (2017). Dysbiosis and its discontents. *mBio* 8, e01492–17.
McFall-Ngai, M., Hadfield, M. G., Bosch, T. C., Carey, H. V., Domazet-Loso, T., Douglas, A. E., Dubilier, N., Eberl, G., Fukami, T., Gilbert, S. F., et al. (2013). Animals in a bacterial world, a new imperative for the life sciences. *Proceedings of the National Academy of Sciences USA* 110, 3229–36.
Sender, R., Fuchs, S., and Milo, R. (2016). Revised estimates for the number of human and bacteria cells in the body. *PLoS Biology* 14, e1002533.
Vujkovic-Cvijin, I., Sklar, J., Jiang, L., Natarajan, L., Knight, R., and Belkaid, Y. (2020). Host variables confound gut microbiota studies of human disease. *Nature* 587, 448–54.

Chapter 2: How to get and keep a microbiome

Adair, K. L. and Douglas, A. E. (2016). Making a microbiome: the many determinants of host-associated microbial community composition. *Current Opinion in Microbiology* 35, 23–9.
Brito, I. L., Gurry, T., Zhao, S., Huang, K., Young, S. K., Shea, T. P., Naisilisili, W., Jenkins, A. P., Jupiter, S. D., Gevers, D., and Alm, E. J. (2019). Transmission of human-associated microbiota along family and social networks. *Nature Microbiology* 4, 964–71.
Byrd, A. L., Belkaid, Y., and Segre, J. A. (2018). The human skin microbiome. *Nature Reviews Microbiology* 16, 143–55.
David, L. A., Maurice, C. F., Carmody, R. N., Gootenberg, D. B., Button, J. E., Wolfe, B. E., Ling, A. V., Devlin, A. S., Varma, Y., Fischbach, M. A., et al. (2014). Diet rapidly and reproducibly alters the human gut microbiome. *Nature* 505, 559–63.
Korpela, K., Helve, O., Kolho, K.-L., Saisto, T., Skogberg, K., Dikareva, E., Stefanovic, V., Salonen, A., Andersson, S., and de Vos, W. M. (2020). Maternal fecal microbiota transplantation in Cesarian-born infants rapidly restores normal gut microbial development. *Cell* 183, 324–34.
Lundberg, D. S., Lebeis, S., Paredes, S. H., Yourstone, S., Ghering, J., Malfatti, S., Tremblay, J., Engelbrektson, A., Kunin, V.,

del Rio, T. G., et al. (2012). Defining the core *Arabidopsis thaliana* root microbiome. *Nature* 488, 86–90.
Sela, D. A. and Mills, D. A. (2010). Nursing our microbiota: molecular linkages between bifidobacteria and milk oligosaccharides. *Trends in Microbiology* 18, 298–307.
Zipfel, C. and Oldroyd, G. E. D. (2018). Plant signalling in symbiosis and immunity. *Nature* 543, 328–36.

Chapter 3: Microbiomes, nutrition, and metabolic health

Ankrah, N. Y. D. and Douglas, A. E. (2018). Nutrient factories: metabolic function of beneficial microorganisms associated with insects. *Environmental Microbiology* 20, 2002–11.
Chaudhari, S. N., Luo, J. N., Harris, D. A., Aliakbarian, H., Yao, L., Paik, D., Subramaniam, R., Adhikari, A. A., Vernon, A. H., Kilic, A., et al. (2021). A microbial metabolite remodels the gut–liver axis following bariatric surgery. *Cell Host Microbe* 10, 408–24.
Fan, Y. and Pedersen, O. (2021). Gut microbiota in human metabolic health and disease. *Nature Reviews Microbiology* 19, 55–71.
Koeth, R. A., Wang, Z., Levison, B. S., Buffa, J. A., Org, E., Sheehy, B. T., Britt, E. B., Fu, X., Wu, Y., Li, L., et al. (2013). Intestinal microbiota metabolism of L-carnitine, a nutrient in red meat, promotes atherosclerosis. *Nature Medicine* 19, 576–85.
Smith, M. I., Yatsunenko, T., Manary, M. J., Trehan, I., Mkakosya, R., Cheng, J., Kau, A. L., Rich, S. S., Concannon, P., Mychaleckyj, J. C., et al. (2013). Gut microbiomes of Malawian twin pairs discordant for kwashiorkor. *Science* 339, 548–54.
Waters, J. L. and Ley, R. E. (2019). The human gut bacteria *Christensenellaceae* are widespread, heritable, and associated with health. *BMC Biology* 17, 83.

Chapter 4: Microbiomes, the brain, and behaviour

Bastiaanssen, T. F. S., Cussotto, S., Claesson, M. J., Clarke, G., Dinan, T. G., and Cryan, J. F. (2020). Gutted! Unravelling the role of the microbiome in major depressive disorder. *Harvard Review of Psychiatry* 28, 26–39.
Cryan, J. F., O'Riordan, K. J., Cowan, C. S. M., Sandhu, K. V., Bastiaanssen, T. F. S., Boehme, M., Codagnone, M. G., Cussotto, S.,

Fulling, C., Golubeva, A. V., et al. (2019). The microbiota-gut-brain axis. *Physiological Reviews* 99, 1877–2013.
Morais, L. H., Schreiber, H. L., and Mazmanian, S. K. (2021). The gut microbiota–brain axis in behaviour and brain disorders. *Nature Reviews Microbiology* 19, 241–55.
Schretter, C. E., Vielmetter, J., Bartos, I., Marka, Z., Marka, S., Argade, S., and Mazmanian, S. K. (2018). A gut microbial factor modulates locomotor behaviour in Drosophila. *Nature* 563, 402–6.
Sharon, G., Cruz, N. J., Kang, D.-W., Gandal, M. J., Wang, B., Kim, Y.-M., Zink, E. M., Casey, C. P., Taylor, B. C., Lane, C. J., et al. (2019). Human gut microbiota from autism spectrum disorder promote behavioral symptoms in mice. *Cell* 177, 1600–18.
Wu, S. C., Cao, Z. S., Chang, K. M., and Juang, J. L. (2017). Intestinal microbial dysbiosis aggravates the progression of Alzheimer's disease in Drosophila. *Nature Communications* 8, 24.

Chapter 5: Microbiomes and infectious disease

Beckmann, J. F., Bonneau, M., Chen, H., Hochstrasser, M., Poinsot, D., Mercot, H., Weill, M., Sicard, M., and Charlat, S. (2019). The toxin-antidote model of cytoplasmic incompatibility: genetics and evolutionary implications. *Trends in Genetics* 35, 175–85.
Daisley, B. A., Chmiel, J. A., Pitek, A. P., Thompson, G. J., and Reid, G. (2021). Missing microbes in bees: how systematic depletion of key symbionts erodes immunity. *Trends in Microbiology* 28, 1010–20.
Kim, S., Covinton, A., and Pamer, E. G. (2017). The intestinal microbiota: antibiotics, colonization resistance, and enteric pathogens. *Immunological Reviews* 279, 90–105.
Ross, P. A., Turelli, M., and Hoffmann, A. A. (2019). Evolutionary ecology of Wolbachia: releases for disease control. *Annual Review of Genetics* 53, 93–116.
Schlatter, D., Kinkel, L., Thomashow, L., Weller, D., and Paulitz, T. (2017). Disease suppressive soils: new insights from the soil microbiome. *Phytopathology* 107, 1284–97.
Stevens, E. J., Bates, K. A., and King, K. C. (2021). Host microbiota can facilitate pathogen infection. *PLoS Pathogens* 17, e1009514.
Vaishnava, S., Yamamoto, M., Severson, K. M., Ruhn, K. A., Ya, X., Koren, O., Ley, R., Wakeland, E. K., and Hooper, L. V. (2011). The antibacterial lectin RegIIIgamma promotes the spatial segregation of microbiota and host in the intestine. *Science* 334, 255–8.

Walker, T., Johnson, P. H., Moreira, L. A., Iturbe-Ormaetxe, I., Frentiu, F. D., McMeniman, C. J., Leong, Y. S., Dong, Y., Axford, J., Kriesner, P., et al. (2011). The *w*Mel Wolbachia strain blocks dengue and invades caged *Aedes aegypti* populations. *Nature* 476, 450–3.

Wischmeyer, P. E., Tang, H., Ren, Y., Bohannon, L., Ramirez, Z. E., Andermann, T. M., Messina, J. A., Sung, J. A., Hensen, D., Jung, S.-H., et al. (2022). Daily *Lactobacillus* probiotic versus placebo in COVID-19-exposed household contacts (PROTECT-EHC): a randomized clinical trial. *MedRχiv*. <https://doi.org/10.1101/2022.01.04.21268275> (posted 22 Jan. 2022)

World Mosquito Program. The use of *Wolbachia*-infected mosquitoes to suppress human viral disease such as dengue, available at <www.worldmosquitoprogram.org>.

Chapter 6: Plant microbiomes in agriculture and food production

Hetherington, A. J. and Dolan, L. (2018). Stepwise and independent origins of roots among land plants. *Nature* 561, 235–8.

Noe, R. and Kiers, E. T. (2018). Mycorrhizal markets, firms, and co-ops. *Trends in Ecology and Evolution* 33, 777–89.

Pankievicz, V. C. S., Irving, T. B., Maia, L. G. S., and Ane, J.-M. (2019). Are we there yet? The long walk towards the development of efficient symbiotic associations between nitrogen-fixing bacteria and non-leguminous crops. *BMC Biology* 17, 99.

Salas-Gonzalez, I., Reyt, G., Flis, P., Custodio, V., Gopaulchan, D., Bakhoum, N., Dew, T. P., Suresh, K., Franke, R. B., Dangle, J. L., et al. (2021). Coordination between microbiota and root endodermis supports plant mineral nutrient homeostasis. *Science* 371, eabd0695.

Song, C., Jin, K., and Raaigmakers, J. M. (2021). Designing a home for beneficial plant microbiomes. *Current Opinion in Plant Biology* 62, 1–10.

Thirkell, T. J., Charters, M. D., Elliott, A. J., Sait, S. M., and Field, K. J. (2017). Are mycorrhizal fungi our sustainable saviours? *Journal of Ecology* 105, 921–9.

Trivedi, P., Leach, J. E., Tringe, S. G., Sa, T., and Singh, B. K. (2020). Plant-microbiome interactions: from community assembly to health. *Nature Reviews Microbiology* 18, 607–21.

Chapter 7: Microbial therapies and healthy microbiomes

Blaser, M. J. and Falkow, S. (2009). What are the consequences of the disappearing human microbiota? *Nature Reviews Microbiology* 7, 887–94.

Buffie, C. G., Bucci, V., Stein, R. R., McKenney, P. T., Ling, L., Gobourne, A., No, D., Liu, H., Kinnebrew, M., Viale, A., et al. (2015). Precision microbiome reconstitution restores bile acid mediated resistance to *Clostridium difficile*. *Nature* 517, 205–8.

Gehrig, J. L., Venkatesh, S., Chang, H.-W., Hibberd, M. C., Kung, V. L., Cheng, J., Chen, R. Y., Subramanian, S., Cowardin, C. A., Meier, M. F., et al. (2019). Effects of microbiota-directed foods in gnotobiotic animals and undernourished children. *Science* 365, eaau4732.

Lin, D. M., Koskella, B., and Lin, H. C. (2017). Phage therapy: an alternative to antibiotics in the age of multi-drug resistance. *World Journal of Gastrointestinal Pharmacology and Therapeutics* 8, 162–73.

Sonnenburg, E. D., Smits, S. A., Tikhonov, M., Higginbottom, S. K., Wingreen, N. S., and Sonnenburg, J. L. (2016). Diet-induced extinctions in the gut microbiota compound over generations. *Nature* 529, 212–15.

Strachan, D. P. (1989). Hay fever, hygiene, and household size. *British Medical Journal* 299, 1259–60.

Suez, J. and Elinav, E. (2017). The path towards microbiome-based metabolite treatment. *Nature Microbiology* 2, 17075.

Trevelline, B. K., Fontaine, S. S., Hartup, B. K., and Kohl, K. D. (2019). Conservation biology needs a microbial renaissance: a call for the consideration of host-associated microbiota in wildlife management practices. *Proceedings of the Royal Society B* 286, 20182448.

Index

For the benefit of digital users, indexed terms that span two pages (e.g., 52–53) may, on occasion, appear on only one of those pages.

G

H

I

K

L

M

N

O

P

R

S

T

V

W

SCIENTIFIC REVOLUTION
A Very Short Introduction
Lawrence M. Principe

In this *Very Short Introduction* Lawrence M. Principe explores the exciting developments in the sciences of the stars (astronomy, astrology, and cosmology), the sciences of earth (geography, geology, hydraulics, pneumatics), the sciences of matter and motion (alchemy, chemistry, kinematics, physics), the sciences of life (medicine, anatomy, biology, zoology), and much more. The story is told from the perspective of the historical characters themselves, emphasizing their background, context, reasoning, and motivations, and dispelling well-worn myths about the history of science.

www.oup.com/vsi

The History of Medicine

A Very Short Introduction

William Bynum

Against the backdrop of unprecedented concern for the future of health care, this Very Short Introduction surveys the history of medicine from classical times to the present. Focussing on the key turning points in the history of Western medicine, such as the advent of hospitals and the rise of experimental medicine, Bill Bynum offers insights into medicine's past, while at the same time engaging with contemporary issues, discoveries, and controversies.

www.oup.com/vsi

EPIDEMIOLOGY

A Very Short Introduction

Rodolfo Saracci

Epidemiology has had an impact on many areas of medicine; from discovering the relationship between tobacco smoking and lung cancer, to the origin and spread of new epidemics. However, it is often poorly understood, largely due to misrepresentations in the media. In this *Very Short Introduction* Rodolfo Saracci dispels some of the myths surrounding the study of epidemiology. He provides a general explanation of the principles behind clinical trials, and explains the nature of basic statistics concerning disease. He also looks at the ethical and political issues related to obtaining and using information concerning patients, and trials involving placebos.

www.oup.com/vsi

Forensic Science

A Very Short Introduction

Jim Fraser

In this Very Short Introduction, Jim Fraser introduces the concept of forensic science and explains how it is used in the investigation of crime. He begins at the crime scene itself, explaining the principles and processes of crime scene management. He explores how forensic scientists work; from the reconstruction of events to laboratory examinations. He considers the techniques they use, such as fingerprinting, and goes on to highlight the immense impact DNA profiling has had. Providing examples from forensic science cases in the UK, US, and other countries, he considers the techniques and challenges faced around the world.

> An admirable alternative to the 'CSI' science fiction juggernaut . . . Fascinating.
>
> **William Darragh, Fortean Times**

www.oup.com/vsi

The History of Life

A Very Short Introduction

Michael J. Benton

There are few stories more remarkable than the evolution of life on earth. This *Very Short Introduction* presents a succinct guide to the key episodes in that story - from the very origins of life four million years ago to the extraordinary diversity of species around the globe today. Beginning with an explanation of the controversies surrounding the birth of life itself, each following chapter tells of a major breakthrough that made new forms of life possible: including sex and multicellularity, hard skeletons, and the move to land. Along the way, we witness the greatest mass extinction, the first forests, the rise of modern ecosystems, and, most recently, conscious humans.

www.oup.com/vsi